# MAKING A DIFFERENCE

## The Fire Officer's Role

**STUDENT TEXTBOOK**

**ONGUARD**
TRAINING FOR LIFE

Fire officer training is not an exact science and it involves many unknown variables. *Making A Difference* is not intended to provide all the training necessary for complete officer training. OnGUARD will not be responsible for any misunderstanding or misapplication of the information presented in this program. In addition, OnGUARD will not be liable for any death or injury resulting from the application of information presented in this course.

**ONGUARD**
TRAINING FOR LIFE

# ACKNOWLEDGMENTS

**Randy R. Bruegman**

OnGUARD would like to acknowledge Randy R. Bruegman, Executive Producer and contributing author for *Making A Difference: The Fire Officer's Role.* Mr. Bruegman is the Fire Chief for the Hoffman Estates, Illinois Fire Department. He has over 20 years of experience as both a paid and a volunteer firefighter. Chief Bruegman holds a Master of Science degree in Management, a Bachelor's degree in Business Administration, and an Associate of Arts degree in Fire Science. He is the Vice Chairman of the International Association of Fire Chiefs Accreditation Committee and serves on the Editorial Advisory Board for *Fire Chief Magazine.*

# INTRODUCTION
# MODULE

"There is nothing more difficult, more perilous,
or more uncertain of success,
than to take the lead in introducing a new order of things."
— Machiavelli

# INTRODUCTION MODULE NOTES

# INTRODUCTION MODULE

## COURSE OBJECTIVES

The objectives of this course are to provide the participants with:

- information on "models" that describe concepts or techniques that are useful to the fire officer
- an assessment of some of their own personal characteristics that effect their behavior as an officer
- some techniques for continued evaluation and improved performance in future assignments
- information on resources that will aid them in fulfilling their duties in the fire service.

## OUTLINE OF COURSE KEY POINTS

During this course, we will be discussing the following topics:

**Tradition** — Anything you keep on doing once you have lost the original reason for doing it. Really, it is our heritage, where we have come from.

**Styles** — There are different leadership and management styles. Each officer needs to know and

**Figure Intro-1   Gear Train**

identify what type of style he/she uses.

**Reality** — We will discuss what goes on in the fire house and the tools you need developed to convert theory to reality. Positions of authority do not necessarily mean competence. Your officer's badge did not come with a set of instructions on how to evaluate, hire, fire, handle yourself in emergency situations, or deal with others in general.

**Reputation** — Anything you can do in 30 seconds that takes a lifetime to live down. A reputation consists of things that people never let you forget. Practicing your profession and a high level of performance can cause your reputation to be changed or built.

**Interpersonal Skills** — Learning to react in a more positive way to your subordinates can cause the entire organization to move from low gear into high gear. This concept is represented by the **Gear Train Model** (Figure Intro-1). Each officer must learn how to drive the organization and not be merely driven by his/her subordinates. Giving out information does not mean that you are getting through to people. What counts is getting them to do the correct things and have those involved feel good about the results.

**Modeling Theory** — This theory is based on the **Input/Output Model** (Figure Intro-2), which tells us that what you put into the process will determine what the final result will be. There are four possible results in a fire house:

a. things can get better,

b. things can stay the same,

c. things can get worse,

d. or no one knows what is going to happen. This creates a fear of the unknown for everyone involved.

**Leaders** — Those who do the right things.

**Managers** — Those who do things in the right way.

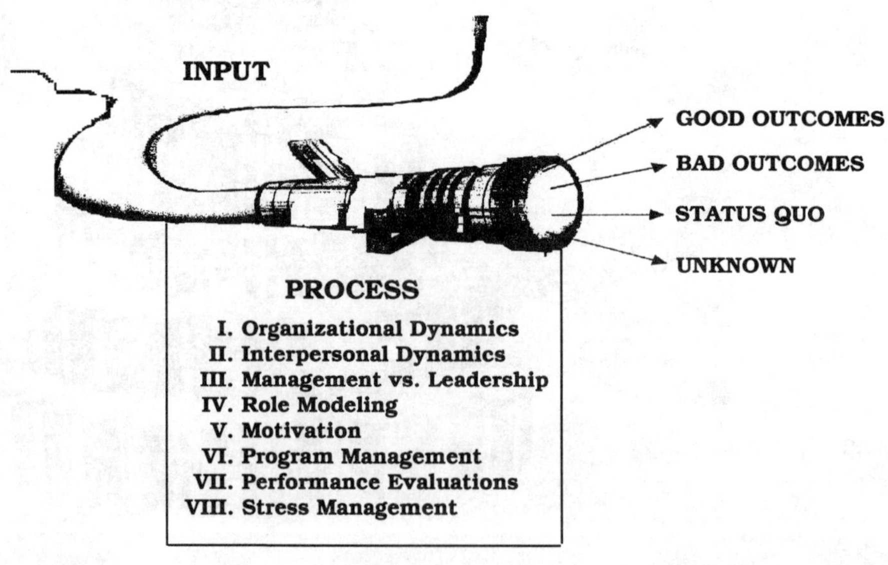


**Figure Intro-2   Input/Output Model**

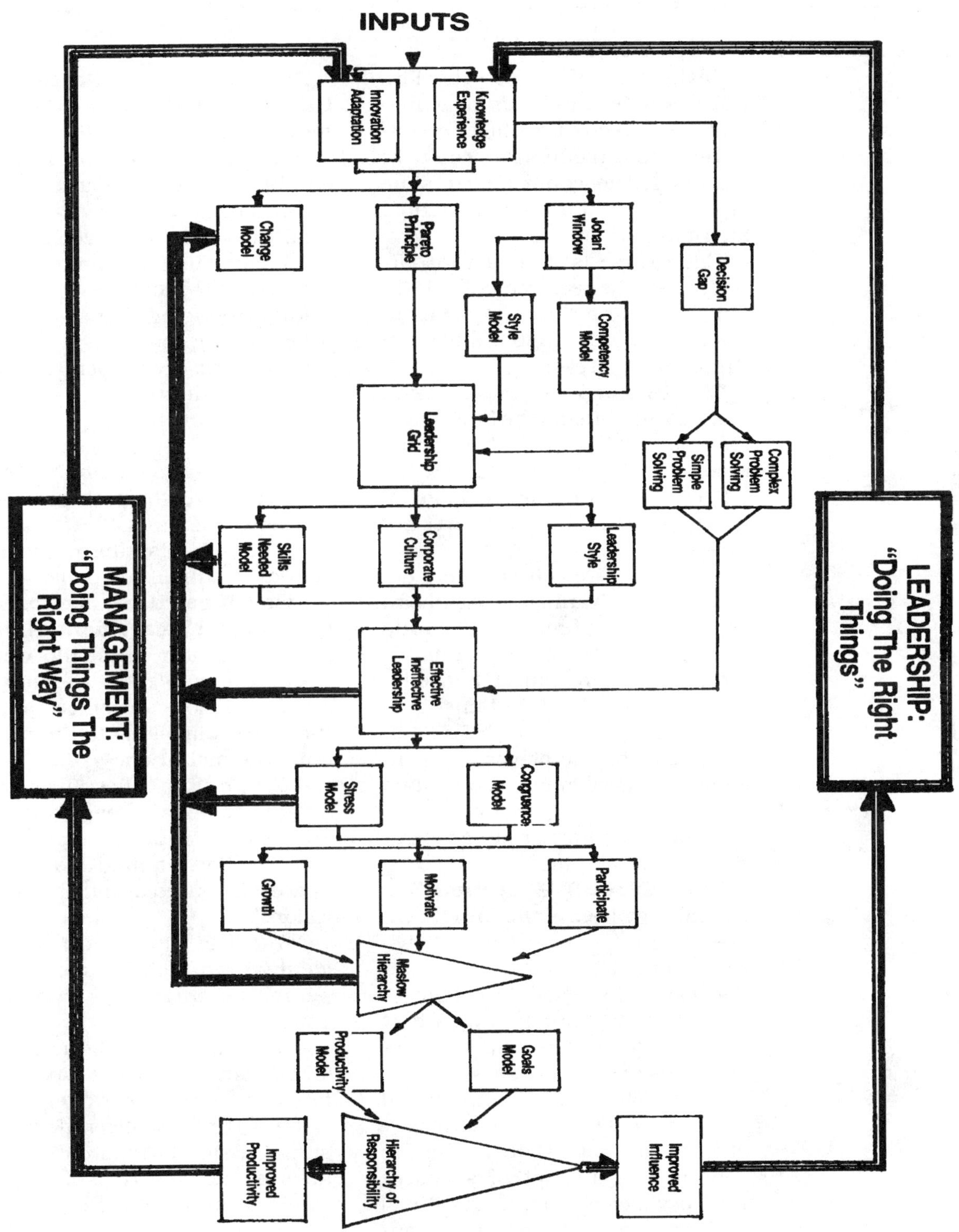


**Figure Intro-3   Systems Approach to Professionalism**

## APPLICABLE NFPA STANDARDS

"Making a Difference: The Fire Officer's Role" can be used to augment many state fire officer certification requirements. You are encouraged to contact your state director of training for additional information.

This course is designed to partially meet the requirements of National Fire Protection Association, Standard 1021 (1986 Edition) in the following areas:

**Fire Officer I —Section A-2-2. Leadership.** Intent: The Fire Officer I should understand the basic processes of management and how the officer applies the principles of leadership to fulfill the supervisory function.

2-2.1 The Fire Officer I shall describe the advantages, disadvantages, and effects of various recognized styles of leadership.

2-2.2 The Fire Officer I, given a summary of the functions of a leader, shall:

(a) describe the officer's responsibility in promoting cooperation and

(b) describe how group cooperation may be obtained

**Fire Officer II — Section 3.1. Psychology (General).** The Fire Officer II shall demonstrate knowledge of the emotional and behavioral characteristics of the individual or working group as they apply to the responsibilities of subordinates and supervisors.

Intent: The Fire Officer II should understand basic principles of psychology as they relate to the fire service. A transcript of a college level psychology course, results of a Clep test, or an equivalent method of testing is acceptable.

**Fire Officer II — Section 3.2. Human Relations and Management.** The Fire Officer II shall describe how each of the following effects the group behavior within the organization:

(a) Understanding people

(b) Motivating the worker

(c) Handling disputes

(d) Introducing changes

(e) Gaining cooperation

(f) Supervisory cooperation

(g) Job attitude

(h) Company policy

(i) Emotional status

(j) Handling complaints

(k) Handling the problem worker

Intent: The Fire Officer II should have an understanding of the basic factors of human relations.

**Fire Officer III — Section 4.7. Personnel Management.** Intent: The Fire Officer III should understand the processes of management and how the officer applies the leadership skills to fire protection administration.

4-7.1 The Fire Officer III, given an actual or simulated personnel situation shall identify the types of corrective action required and describe how each is used as a tool of supervision.

4-7.2 The Fire Officer III shall identify and define three types of personnel interviews commonly used.

4-7.3 The Fire Officer III shall identify and define two types of approaches to counseling.

4-7.4 The Fire Officer III, given an actual or simulated situation, shall determine which type of interview would be most valuable.

4

## THE TRUMPETS THAT WE WEAR

These days you hear a lot about professionalism. There is a fair amount of time spent discussing what professionalism is and how it can be achieved by the fire service. No one seems to have come up with a definition of professionalism that is acceptable to everyone.

There are examples that people refer to as showing professionalism, such as more education, getting paid to do a service, and subtle changes in image. Currently one trend that is sweeping the fire service is to combine police departments with fire departments to form what is commonly called "public safety departments." However, we may be throwing the baby out with the proverbial bath water!

If we lose sight of our identity, then perhaps we will also lose sight of what it is we are supposed to stand for. There is a question we need to ask, "Are we truly professionals?" Perhaps it is time for us to examine the symbolism, tradition, and ethics associated with fulfilling the role of the fire officer before we decide about our level of professionalism. Let's start with the symbol of our rank — the fire trumpet.

### A Symbol Is a Promise

When you became a fire officer someone gave you a badge and a set of collar ornaments that probably had one or more scaled down speaking trumpets inscribed upon the metal. Those trumpets are symbolic of the role of the officer. They are a symbol that has become part of the fire service's "tradition." The whole idea came from the days when the symbol of command was the speaking trumpet. Have you ever wondered why we keep adding trumpets to the cluster as you move up in rank? Why still use the trumpet as a symbol? Why do we stop at only five trumpets?

The speaking trumpet is to the fire service what the caduceus is to the medical profession. It is our symbol of professionalism and should be displayed frequently as a reminder of the various roles that a modern fire officer has in the pursuit of fire protection. After working so hard to earn those trumpets we should wear them proudly as a symbol of our rank.

I previously asked a question about the number of trumpets on a badge. Just for the sake of argument, I'll suggest that there are five because a fire chief, the ultimate fire officer, has five basic roles to fulfill in order to advance the profession. These roles are:

- leader
- manager
- role model
- mentor
- change agent

As firefighters promote to the rank of lieutenant or captain, they also assume each of these roles.

These roles impose a burden of responsibility that must be balanced almost daily if an officer is to succeed in his/her role. One role cannot be emphasized to the detriment of the other without harming the organization. Yet, each role is a separate function that creates unique demands upon an officer.

## Leadership — The First Trumpet

The first trumpet we wear stands for *leadership*. The role of a leader is to be out in front, take risks, and be visible. In the old days, the fire officer was truly a leader in this sense. He was in the forefront of the battle against fire. Today's demands on the fire officer are increasingly complex. Conducting fire inspections, performing public education, preparing for hazardous materials accidents, assisting in company training, and preparing departmental budgets are just a few of today's battles. The battlefield has changed greatly, but the role of leadership remains the same.

The leadership trumpet continues to impose a great burden on a fire officer. An officer is expected to not only exercise leadership, but also to take charge in the field of fire protection. It imposes a burden of obligation to get out in front of change instead of reacting to it. That's what leadership is all about — taking someone, or a group of "someones," somewhere.

At the end of the 1800's, there was a fire chief in New York by the name of James Gulick. Some called him brash and some called him flamboyant, but all of his men called him Chief. Once, when a group of politicians tried to take action that would have devastated the fire department, he took a stand. Facing the political opposition on the steps of the city hall in New York, he removed his helmet and told the politicians that if they did not listen to him on the issue he would resign. The politicians called his bluff and he called theirs.

When "Gentleman Jim's" helmet hit the ground, it was echoed by thousands of volunteer firefighters' helmets clattering to the ground in unison. The issue was soon resolved in favor of the fire service and James Gulick finished out his career without another serious challenge. It took courage to **lead** a group of men under such desperate conditions.

Today leadership is at a premium because the process of "appointing" a fire officer often reveals large differences between the supervisors and those being supervised. Nevertheless, firefighters want to feel that their officer is a leader, a person who will not fail them in a crisis.

> *An officer must make sure that the marriage of delivering fire protection to the community is matched with a keen sense of cost effectiveness and cost benefit.*

## Manager — The Second Trumpet

The second trumpet symbolizes the role of a *manager,* which is also difficult to master, but a necessary part of the formula. Fire protection is costly and can be administered either effectively or ineffectively. This role includes fiscal responsibility. In other words, an officer must make sure that the marriage of delivering fire protection to the community is matched with a keen sense of cost effectiveness and cost benefit.

Much of the credibility of a fire officer comes from their ability to predict what effects there will be from a certain expenditure of funds. Managing means manipulating resources to achieve predictable results. It means managing manpower to achieve productivity. It means knowing both sides of the ledger in the field of fire protection — expenditures and revenues. A modern fire officer must know and utilize a wide variety of facts in the delivery of the fire protection services. Managing means to being top of things.

Someone once said that ***managing*** an organization meant that you do things right. The same person also stated that ***leading*** an organization meant that you do the right thing. There is a subtle, but important, distinction. Leaders take people to places they have never been, while managers make sure they get there on time.

When an officer receives the next two trumpets, he/she assumes roles that are difficult to define and even more difficult to assess — those of role model and mentor. Being in a leadership role has a price to pay. One of the first dues it demands is that an officer can no longer avoid responsibility for their actions. As officers ascend the hierarchy of rank, they become more visible to two groups of people, namely those in their organization and those serving in similar organizations.

*Leaders take people to places they have never been, while managers make sure they get there on time.*

## Role Model — The Third Trumpet

The third trumpet in the fire officer's pentacluster represents the task of becoming a ***role model*** for others to follow. In short, one of the tasks of the officer is to set the example. If we ever hope to achieve excellence in the profession, we have to strive for it ourselves. There is a lot of truth in the cliche that "seeing is believing." If we don't strive for goals ourselves, we cannot expect our subordinates to reach for higher goals. If we do not demand competency in our everyday actions, we cannot expect competency from those who are looking to us as an example.

A role model may often be a person that is distant or remote. A role model can be living or dead. A role model provides an image that can be used to map behavior. One of my early role models was a chief by the name of Keith Klinger. I never actually met Chief Klinger until I was promoted to chief myself, but I tried to pattern my professional life after his.

Role models are like that. In today's fire service we have several role models, such as Fire Chief Alan Brunacini of Phoenix, Arizona. Role models are often larger than life and seem unapproachable. However, in reality they provide a blueprint that causes others to say, "I want to be like that person." They are someone you want to be **like**, not the **same** as. They are someone you can respect, a person you can emulate in style and philosophy. If you are to be a role model, you must exhibit behavior worthy of being emulated.

*If we ever hope to achieve excellence in the profession, we have to strive for it ourselves.*

## Mentoring — The Fourth Trumpet

Setting the example is not enough. The fourth trumpet requires that we counsel others that are following in our footsteps. This is called ***mentoring.*** Officers have to overcome obstacles to advance in their career. In the process, they obtain a great deal of information that cannot be found between the pages of a textbook or in lecture notes at college. In order to elevate the profession, we need to share our experience and information in an informal way. We should find a younger version of ourselves and pass this information on to them.

To mentor is to guide. A mentor is someone that can help another person solve a problem, but who will not attempt to solve it for them.

Eventually the student should be able to listen and begin to think like their mentor.

Being a mentor means to risk helping a person grow at a more rapid pace than you have been able to grow. To mentor is to risk creating a replacement that may have more to contribute than you do. To mentor also means to achieve a form of immortality. You have a unique opportunity to become a little bit of the future because you have invested in a person who will actually help shape that future.

Don't mistake the role of mentor for friend. My first mentor was a man that made me stretch to the limit at times. I often felt less than grateful, but I always felt I had grown in the process. This particular individual was a company officer that I actually bypassed later in the promotional process. However, I never lost the feeling that I could call upon him for advice. Since that first mentor experience, I have had several others. Each has helped me to grow in a different way and always in new directions.

To summarize the role model and mentor relationship is simple. Once we have assumed the mantle of responsibility that comes with being a fire officer, we need to become role models for others to follow. We must be willing to share what we know with those that are following in our footsteps. If we don't, then those following us will be traveling a path without signposts to lead them to future success. We owe it to our subordinates and profession to mark the path to success clearly.

> *To mentor means to achieve a form of immortality by investing in a person who will help shape the future.*

## Change Agent — The Fifth Trumpet

The fifth trumpet that overlays the other roles is that of **change agent**. A change agent is a person that makes things happen. Someone once told me that there were only three kinds of people in the world:

- those that "watched things happen"
- those that "made things happen"
- and those that "wondered what happened."

If a person has grown into the other roles symbolized by trumpets on the way up the ladder, then the only task left is to use the last trumpet to make things better than they were when you were aspiring for the rank.

The ultimate role of the fire officer is to leave a **mark** on the system, not just a memory. There is a tradition for the fire officer that we have not mentioned. That tradition is that the officer leaves a legacy of improvements during their era in power. As a former combat officer once stated, "No guts, no glory." The change agent takes a risk in pushing back the darkness of limited knowledge and reducing the prejudice against change. If the officer doesn't take this risk, then his role is unfulfilled.

Change is called the "cutting edge" because it represents the potential for trauma and injury. We, on the other hand, know that failure to change leads to obsolescence. The fifth trumpet weighs the heaviest of all because it puts a burden on the officer to deal with the collective resistance to change. Officers of all ranks in the fire service must be ready to challenge the status quo. The officer must be the one ready to question the validity of programs and actions of the organization. Lastly, the officer must be the one to make sure that someone takes action on needed changes.

Not everyone wants the burden of accepting the role of change agent. Unfortunately, not everyone has the ability either. Those who choose to accept the challenge have a tremendous opportunity to stand in the footsteps of all their predecessors that have brought us to where we are today.

In this world of ever changing definitions, the terms we use to identify some jobs as professional are sort of meaningless. Janitors are called professional "sanitation engineers," plumbers also call their work a profession. The real measure of a profession is not found in what the profession calls itself, but rather in how a member of the profession acts and produces. The medical profession has an ancient oath and a code of ethics to match. You have to "pass the bar" to practice law in almost every court in the world. Perhaps in the future the fire service will develop a code of conduct and ethical guidelines that will rival other professions.

In light of modern technology, the speaking trumpet is an antique, but what it stands for supersedes the technology of the time. It is the symbol of responsibility. It symbolizes authority that needs to be respected and carried out with dignity.

## Making A Difference

If you are already an officer, the next time you put on the cluster of trumpets, think about the roles that each one stands for. If you aspire to become a fire chief, you must seriously consider whether you want to carry out the roles each trumpet stands for. To accept this challenge will mean that the fulfillment of these roles will aid the fire service in it's mission and contribute to the development of a professional image.

In this course, "Making A Difference: The Fire Officer's Role," our goal is two-fold. First, if you are already a fire officer, we will focus on the unique aspects of fire service management that will help you perform your job more effectively.

Secondly, if you are a firefighter with aspirations of becoming a fire officer, or a fire officer wanting to be a chief, we will provide a base of knowledge and a framework for you to achieve your goals.

Fire officers face a unique set of circumstances seldom encountered by other managers. The stress generated by emergency response, personnel problems resulting from living with each other for extended periods of time, and the blending of tradition with rapid changing technologies are just a few of the unique challenges facing today's fire officers.

There is a phrase on the following page that reflects the intent of this program. Now, if you are ready, turn the page and let's start "Making a Difference."

# DON'T SEARCH FOR EXCELLENCE — CREATE IT!

# MODULE I

## GETTING ALONG AND GETTING THROUGH:

## INTERPERSONAL DYNAMICS

**"The most vital spot in management is the point of contact between worker and boss."**
**— Lawrence A. Appley**

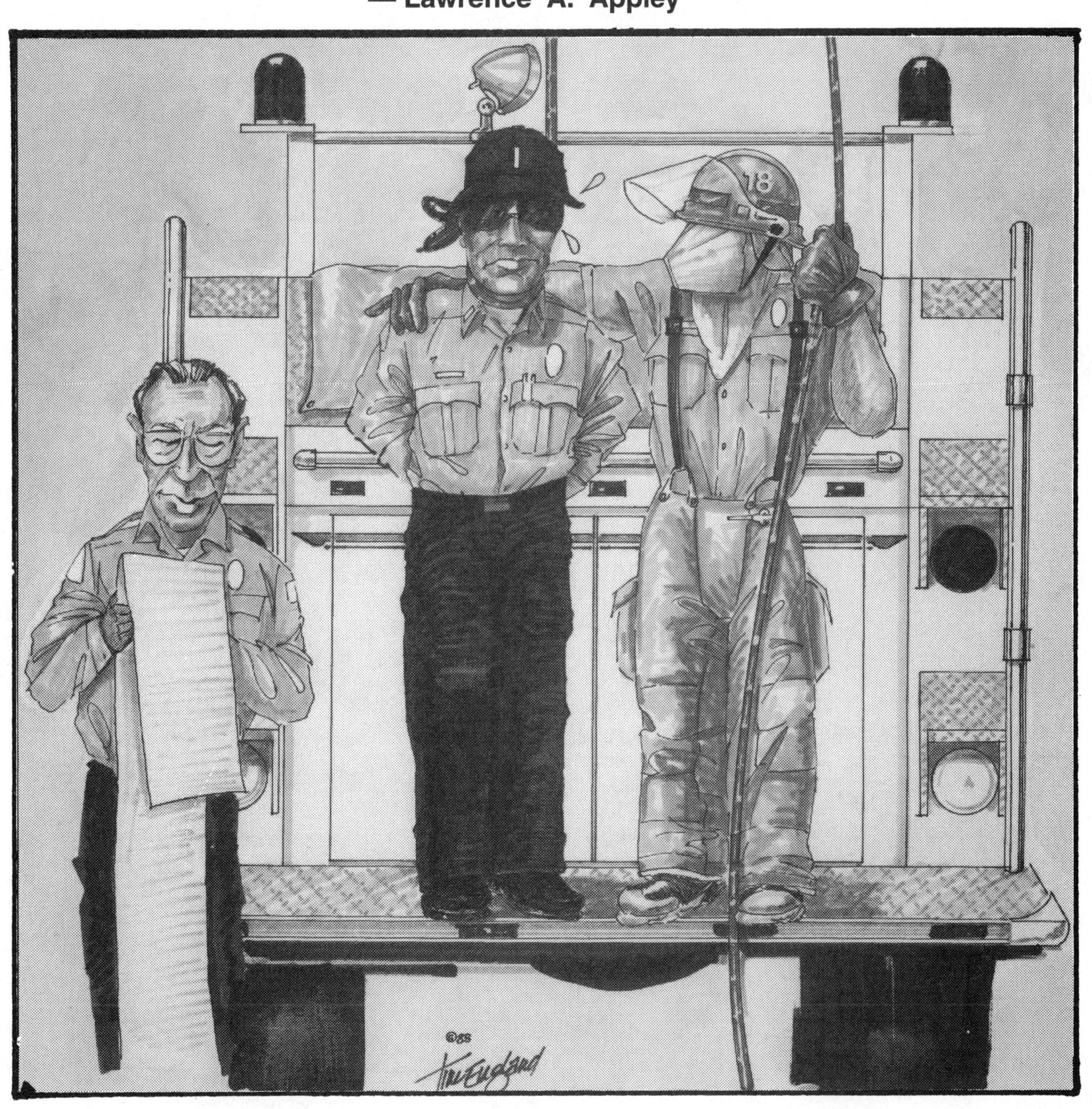

# PRE-PROGRAM ACTIVITY 1

## YOUR COMMUNICATION I.Q.

*Instructions:* Answer the following communication questions "Yes" or "No" by placing an "x" on the appropriate line.

|  | **YES** | **NO** |
|---|---|---|
| 1. Do your subordinates do what you have requested? | _____ | _____ |
| 2. Do you understand what your superiors expect of you? | _____ | _____ |
| 3. Do you listen more than you speak? | _____ | _____ |
| 4. Do you ask for clarification if you don't understand? | _____ | _____ |
| 5. Do you look a person in the eye when speaking to them? | _____ | _____ |
| 6. Do you understand the implications of "body language?" | _____ | _____ |
| 7. Are your communications timed to benefit the receiver? | _____ | _____ |
| 8. Is compromise an essential component of communication? | _____ | _____ |
| 9. Do you respond quickly to requests for action from your superior and/or subordinates? | _____ | _____ |
| 10. Is your vocabulary adequate? | _____ | _____ |
| 11. Do you consciously attempt to improve your vocabulary? | _____ | _____ |
| 12. Do your written messages clearly state the action required? | _____ | _____ |
| 13. Are your verbal communications conducted in a positive manner? | _____ | _____ |
| 14. Are your professional communications based on common experience and knowledge with the receiver? | _____ | _____ |

**TOTAL**      **YES** _____      **NO** _____

# PRE-PROGRAM ACTIVITY 2

## THINKING ABOUT MENTAL HABITS: KEIRSEY TEMPERAMENT SORT[1]

**Part I**

*Instructions:* Circle the statements which best describe you, **E** or **I**:

|  **E** | **I** |
|---|---|
| **E** likes action and variety | **I** likes quiet and time to consider |
| **E** likes to do mental work by talking to people | **I** likes to do mental work privately before talking |
| **E** acts quickly, sometimes without much reflection | **I** may be slow to try something without understanding it first |
| **E** likes to see how other people do a job and see results | **I** likes to understand the idea and work alone or with a few people |
| **E** wants to know what others expect | **I** wants to set his or her own standards |

Count the number of E's and I's you have circled on each side. Enter the letter which has the highest number circled on the line provided: ________

(If you have an equal number of each letter circled, then enter both letters with a slash separating them on the line, e.g. E/I.)

---

[1] Keirsey, David & Bates, Marilyn. *Please Understand Me: Character & Temperament Types,* (c) 1984. Used by permission of David W. Keirsey, Del Mar, CA.

## THINKING ABOUT MENTAL HABITS (CONT.)

**Part II**

*Instructions:* Circle the statements which best describe you, **S** or **N**:

| S | N |
|---|---|
| **S** pays most attention to experience as it is | **N** pays most attention to the facts and how they fit together |
| **S** likes to use eyes, ears, and other senses to find out what's happening | **N** likes to use imagination to come up with new ways to do things |
| **S** dislikes new problems unless there are standard ways to solve them | **N** likes solving new problems, dislikes doing the same thing over |
| **S** enjoys using skills already learned more than learning new ones | **N** likes using new skills and practicing old ones |
| **S** is patient with details but impatient with complicated situations | **N** is impatient with details but doesn't mind complicated situations |

Count the number of S's and N's you have circled on each side. Enter the letter which has the highest number circled on the line provided: __________

(If you have an equal number of each letter circled, then enter both letters with a slash separating them on the line, e.g. S/N.)

## THINKING ABOUT MENTAL HABITS (CONT.)

**Part III**

*Instructions:* Circle the statements which best describe you, **T** or **F**:

|  **T** | **F** |
| --- | --- |
| **T** likes to decide things logically | **F** likes to decide things with personal feelings and values even if they aren't logical |
| **T** wants to be treated with justice and fair play | **F** likes praise and pleasing people, even in unimportant things |
| **T** may neglect and hurt others feelings without knowing | **F** is aware of other feelings |
| **T** gives more attention to ideas or things than relationships | **F** can predict how others will feel |
| **T** doesn't need harmony | **F** gets upset with conflict, values harmony |

Count the number of T's and F's you have circled on each side. Enter the letter with the highest number circled on the line provided: __________

(If you have an equal number of each letter circled, then enter both letters with a slash separating them on the line, e.g. T/F.)

## THINKING ABOUT MENTAL HABITS (CONT.)

**Part IV**

*Instructions:* Circle the statement which best describe you, **J** or **P**:

|  | **J** |  | **P** |
|---|---|---|---|
| **J** | likes to have a plan that is settled and decided in advance | **P** | likes to stay flexible and avoid fixed plans |
| **J** | tries to make things come out the way they "ought to be" | **P** | deals easily with unplanned and unexpected happenings |
| **J** | likes to finish one project before starting another | **P** | starts many projects but may have trouble finishing them all |
| **J** | usually has mind made up | **P** | usually looking for new information |
| **J** | may decide things too quickly | **P** | may decide things too slowly |
| **J** | wants to be right | **P** | wants to miss nothing |
| **J** | lives by standards and schedules that are not easily changed | **P** | lives by making changes to deal with problems as they come along |

Count the number of J's and P's you have circled on each side. Enter the letter with the highest number circled on the line provided: __________

(If you have an equal number of each letter circled, then enter both letters with a slash separating them on the line, e.g. J/P.)

Now, enter your letter from each exercise in the spaces below:

__________  __________  __________  __________

      E/I        S/N        T/F        J/P

# MODULE I NOTES

# MODULE I

# GETTING ALONG AND GETTING THROUGH: INTERPERSONAL DYNAMICS

## OBJECTIVES

To provide the viewer and participant with:

- a basic understanding of the factors involved in human relations
- an assessment of their behavior as an individual
- a basic understanding of the importance of differences in orientation and style in relationship to interpersonal dynamics

## OUTLINE OF KEY POINTS

**Pareto Principle** — This principle expresses the relationship between the vital few and the trivial many. Eighty percent of our events, activities, resources, occurrences, and inputs are generated by twenty percent of the population. This productive group is known as the **vital few.** The reverse is also true. The remaining eighty percent, or the **trivial many**, of the population produces only twenty percent of the output. We need to continually gain new knowledge to make sure we are part of the vital few.

**Johari Window** — This is an important model because it relates to our level of self-awareness and interaction with others, or **interpersonal relationships**. Each of the four sections of the Johari Window has three elements:

- the things we know about ourselves
- the things that are known to others about us

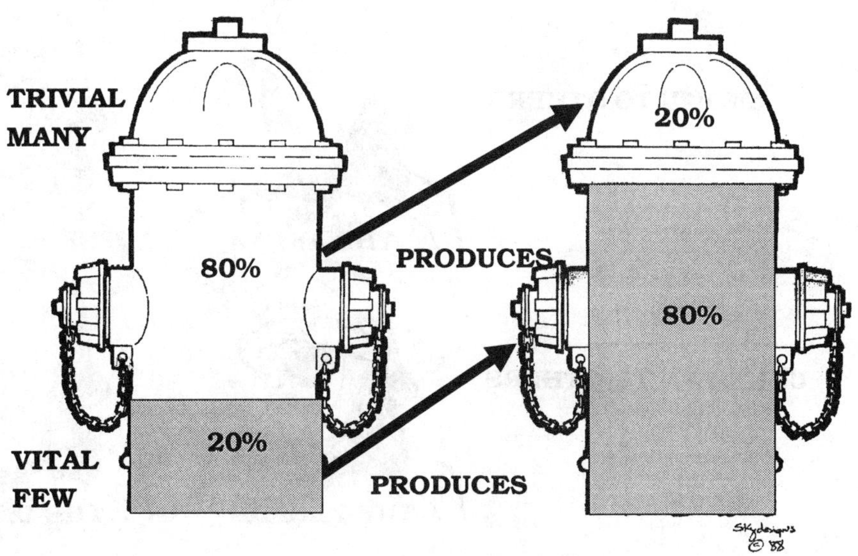


**Figure 1-1   Pareto Principle**

- a combination of the two

The four sections of the Johari Window are:

- **Arena** — The portion of our personality that is "public knowledge." It is what we know about ourselves and what others know about us that we agree upon. The Arena is controlled by us to the degree that we are willing to share information about ourselves with others. The Arena is, in essence, our personality as a whole. When we modify the Arena, we change the information in all the other three areas. This is called "expanding the Arena." It is accomplished by listening to what others see in us and changing ourselves. This requires **listening** to feedback from others.

- **Facade** — The part of our personality that we prevent others from knowing. The knowledge about ourselves that we are unwilling to share with others. The Facade is the part of us where we keep all of our secrets.

- **Blind Spot** — The part of our personality that has an effect on others and yet we are oblivious to it. Blind spots typically are a source of conflict between ourselves and others. It is the portion of our personality that others are aware of, but we are not.

- **Unknown** — The part of our personality that has not been brought to the surface. Our traits that have yet to be uncovered or understood. It is unknown to ourselves and others.

**Exposure** — Our willingness to share our feelings with others.

**Feedback** — The willingness of others to share their feelings with us.

**Figure 1-2  Johari Window**

18

**The Knowledge/Experience Ratio** — The relationship between theory and real-life implications. This model expresses how our level of knowledge and experience effect our decision making process. As knowledge and experience increase, we move from being an amateur to a professional.

- **Amateur** — One who has a limited ability to perform. Ability to perform will increase proportionally with experience. Level of knowledge can be adjusted by training and education.
- **Professional** — Knowledge and experience have increased, dramatically improving the ability to perform. The ratio between knowledge and experience will change with the task or the difficulty of the decision being made. A professional is one who recognizes what they do not know.

**You're Not One of the Boys Anymore** — Once you are an officer, your subordinates rely upon you to make decisions that effect their lives. These decisions vary from daily operations to life-threatening emergency situations.

**Competency Model** — Examines the level of competency and understanding of your own abilities in four types of individuals. It is best to be in a Teacher/Student mode. This individual will teach skills to others, while gaining new knowledge. The four modes in the Competency Model are:

- **The Natural** — An individual who has an innate talent to perform a specific function or skill, but lacks the ability to communicate how to achieve this skill to others. This person is unconsciously competent. They know how to accomplish a given task, but are unaware of how they are able to do it. Their only contribution to the organization is their natural skill.

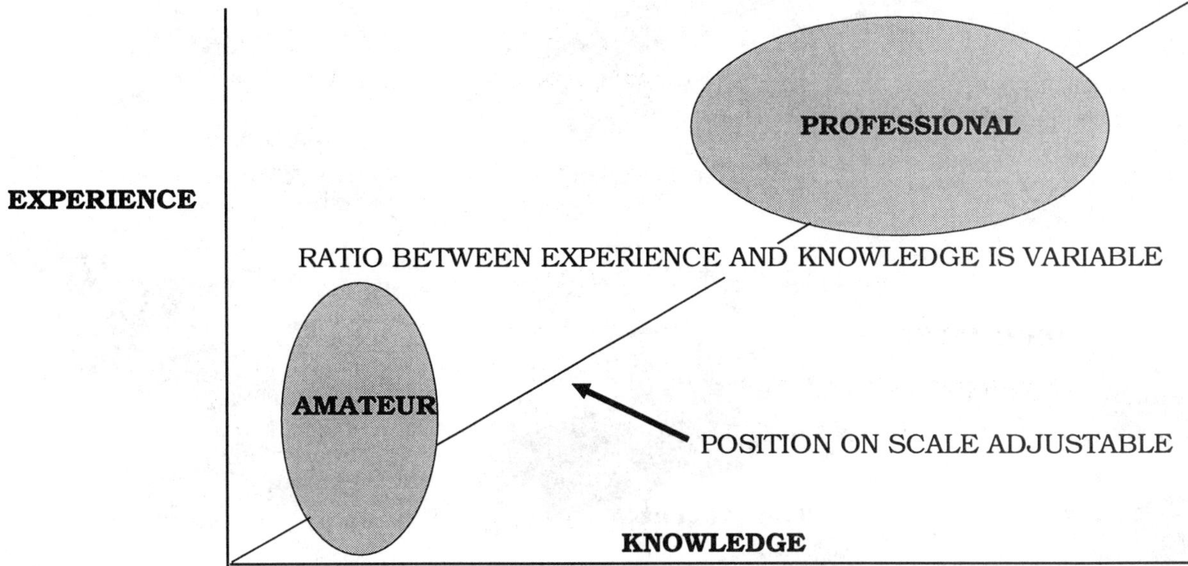


**Figure 1-3   Knowledge/Experience Ratio**

- **The Teacher** — An individual who can perform a specific task with skill and has the ability to communicate to another person how to achieve the same level of competency. This person is consciously competent. They provide direction in an organization which causes things to get better.
- **The Student** — An individual that possesses the ability to perform, but who lacks the training and education to achieve a specific task or accomplishment. A student is a good place to be because they are incompetent and they know it.
- **The Clown** — An individual who lacks an awareness of their own personal skill level. The clown makes minimum contributions to the organization. This person is unconsciously incompetent. The clown doesn't recognize his/her limitations and can be lethal in emergency situations.

**Concept of Morale** — The relationship between an individual and the group, or a group and its organization. Morale results from understanding and communication. It is more than just "feeling good."

**Self-Awareness** — "Ninety percent of the world's woe comes from people not knowing themselves, their abilities, their frailties, and even their real virtues. Most of us go almost all the way through life as complete strangers to ourselves — so how can we know anyone else?" — Sydney J. Harris

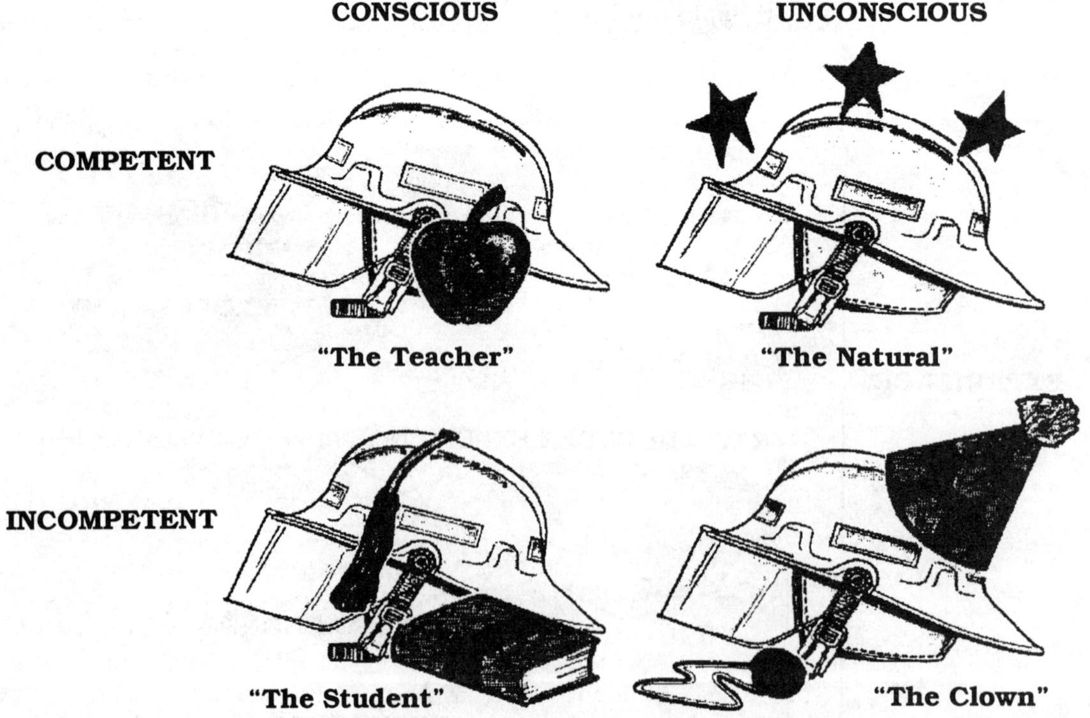


**Figure 1-4   Competency Model**

## WHICH LEADERSHIP STYLES DO YOU USE?

**Extroversion** — Relates more easily to the outer world of people and things than to the inner world of ideas. Focuses on what is going on in the world around them and other people. This type of person is usually more open and direct.

**Introversion** — Relates more easily to the inner world of ideas than to the outer world of people and things. Focuses more on what is going on inside of themselves. This individual is usually more closed and indirect.

**Sensing** — Prefers to work with facts rather than looking for new possibilities and relationships. Thinks in black and white terms.

**Intuitive** — Prefers looking for possibilities and relationships as opposed to working with facts. Oriented toward options and alternatives.

**Thinking** — Bases judgements on impersonal analysis and logic rather than personal values. Logical aspects are more important than the effect on people. Intellectual approach. Asks, "Does it break a rule?"

**Feeling** — Basing judgements more on personal values than on impersonal analysis and logic. Will look at the impact their decision will have on other people.

**Judging** — A preference to a planned, decided, orderly way of life as opposed to a flexible, spontaneous one. These individuals are quite predictable. Will use structured methods to come to a decision.

**Perceiving** — A preference to a flexible, spontaneous way of life as opposed to a planned, decided, orderly one. This individual will look for opportunities. They will question whether the structure needs to be changed.

## POINTS TO REMEMBER ABOUT LEADERSHIP STYLE

*Each organization needs to recognize the strengths and weaknesses of each style and how they can reinforce each other.*

**Recognizing Differences in Style** — Differences of style cause us to look at the world from different perspectives. We need to recognize that these perspectives alter our relationships from time to time. Each organization needs to recognize the strengths and weaknesses of each style and how they can reinforce each other.

**Combination of Style** — Everyone possesses all of the styles within themselves. It is the strength with which we apply a particular style in our everyday lives that causes us to develop a "personality." There are sixteen different combinations of style. Each organization needs at least one of each style. The organization must learn the interaction of each style and how they can compliment one another in order to maximize the group's potential.

**Contrast of Style** — Failure to respect differences in style can result in conflict and miscommunication. Firefighting needs a lot of structure because of the danger to lives if structure is not present in emergency situations. However, there also needs to be a periodic evaluation of the organization to keep it in check.

**Contributions of Style** — Each style has strengths and weaknesses. The combination of styles often brings about a stronger relationship. It can also improve team performance if respect is demonstrated for the contributions of each type.

## INTERPERSONAL DYNAMICS

Just why does a fire officer need to know about communication? What do communication skills have to do with your job? EVERYTHING! Think about how much time you spend preparing to fight fire. You probably spend many hours each month training in various topics like strategy and tactics, ventilation, rescue, etc.

Realistically, how much of your time do you spend fighting fire? Maybe five percent, if your department is like most. The rest of your time is spent around the station, inspecting, and meeting with other firefighters. In short, communicating! Unfortunately, few officers have the training in communication they need. How do you go about improving your communication skills? Read on, and you'll find out.

The term "communication skills" implies that there are effective and ineffective ways to communicate. Not only can you learn the effective methods of communication, but you can improve the way you communicate by practicing these "skills." At first you may feel awkward communicating in ways with which you are not familiar. Soon, when the results of communicating more effectively are apparent, you will be glad you made the effort.

On the surface communication looks easy. It is simply a matter of one person sending a message to another. The sender transmits a message either verbally or nonverbally. The degree to which the receiver understands the original intent of the message depends on many things. We will examine some of these factors in detail.

*Not only can you learn the effective methods of communication, but you can improve the way you communicate by practicing these "skills."*

### Sending Messages

One obvious factor in communication is the environment in which the communication is taking place. Certain situations facilitate communication. Some others get in the way of effective communication. Think about the last "conversation" you had while wearing SCBA masks. Even though such communication is difficult, it is necessary in your job. You **must** tell the other firefighter that you are leaving him for a moment to search the room or direct him to apply the fire stream in a certain spot. Of course, this is an extreme example of communicating in a difficult environment.

Contrast this example with the last time you corrected a subordinate. Was the environment conducive to him hearing your message?

Were you alone, unhurried, calm, and clear? Or were you out on the apparatus floor, rolling hose, or involved in some other activity? Environment does have an effect on communication. There are many other factors which may have even more of an impact.

The sender of the message has a great deal to do with how the message is understood. Being genuine in your communication avoids sending "mixed messages." As a message sender, consider your motives. Are you informing, defending, confronting, reprimanding, praising, or instructing? Each of these activities demands a different style of presentation. The important thing to keep in your mind is **what**

**do I want to accomplish** with this communication.

Each situation is going to be unique. Be clear about why you are communicating and what it is you want to express. If these are clear in your mind, you will improve the chances of having your message understood. Just like speaking on the radio, knowing what you are going to say will increase the chances of your message being understood. Besides being aware of your motives, there are some specific skills that you can master which will help you send clearer messages.

The **dimension of concreteness** refers to speaking specifically to the topic. Try to avoid abstract generalities whenever possible. Give specific examples, be clear in labeling emotions. Being more concrete means being less ambiguous. The more you practice being concrete, the more your messages will be easily understood. Being concrete is important in daily interactions with people, but it is critical when you are giving feedback to someone.

*Being concrete is important in daily interactions with people, but it is critical when you are giving feedback to someone.*

## Giving Feedback

Knowing how to give **feedback** is one of the most basic communication skills. Mastering this skill will improve your ability to supervise people. Many people confuse the words "advice" and "feedback." Consequently, they get into trouble when asked for either. Notice the words "asked for." Feedback should be solicited. Feedback is not always requested when it might be negative. How does a supervisor give feedback, even if it isn't asked for?

Create the kind of working relationship in which feedback is expected, encouraged, and rewarded. Tell your supervisees that it is part of your job to let them know how they are doing. Solicit feedback from them about your performance. Teach them through modeling how to give and receive feedback!

There are two types of feedback — **positive** and **negative**. You must be comfortable in giving both. Your management "bag of tricks" is limited if you only give positive feedback. Sometimes an employee needs firm, negative feedback. You will learn how to do both.

First, you need some rules for giving feedback. For feedback to be effective in changing behavior it must be:

1) About **observable,** changeable behaviors. People need to know how something they did affects their performance.

2) **Descriptive**, rather than evaluative. Describe the result of an action — how it helped or hindered. Avoid judging the person. Simply observe and report.

3) **Specific**, rather than general. Remember to be concrete in your description. Tell how you felt about an action.

4) **Timely**. It does no good to praise or criticize something that the person doesn't remember. Avoid "saving up" too many negatives, so the worker won't feel picked upon. Practice giving positive feedback immediately. Prompt rewards are the surest way to reinforce appropriate behaviors.

OK, now you know the rules. Let's see how they work in practice. Have some fun picking out the pros and cons in the following feedback examples:

### POSITIVE NON-FACILITATIVE FEEDBACK

**Sender**: "By the way, you really did well with that patient last shift."

**Receiver**: "Gee, thanks."

### NEGATIVE NON-FACILITATIVE FEEDBACK

**Sender**: "You really chickened out on that last fire, Bob."

**Receiver**: "Oh yeah? Well I don't think so."

Notice that in both cases, the receiver doesn't know what he did to deserve the feedback. The effect of such feedback is likely to be confusion.

### POSITIVE FACILITATIVE FEEDBACK

**Sender**: "I'm pleased with your treatment of that last patient. The way you told her just what you were doing helped her to calm down."

**Receiver**: "Thanks. I have been trying to work on giving the patients more information. I think it does help them."

### NEGATIVE FACILITATIVE FEEDBACK

**Sender**: "I am angry with you because you hesitated before searching that bedroom."

**Receiver**: "I didn't hesitate. My air tank got stuck on the door knob, and I couldn't go either way. Your pushing only made me more stuck."

Notice in these examples how communication is facilitated. In both cases the people involved know more about each other than they did before. Another principle of feedback is that a receiver of feedback is in charge of what he does with that information. Recall the proverb, "If one man calls you an ass, tell him to go to hell. If five people call you an ass, go buy a saddle." You will have a chance to practice giving feedback as a student activity. It is one of the most important skills involved in sending messages. Next we will consider "receiving" skills.

## Receiving Messages

Equally important as skills involved in sending messages are those involved with receiving them. Once again, it helps to know who and what you are. The best attribute you can have is a solid understanding of yourself. Picture the last time your supervisor chewed you out. He came in, obviously upset, and proceeded to really let you have it. How did you react? Were you defensive, perhaps embarrassed? Did you buy into his perception of you? Hopefully, you were able to realize what was really going on. This is an insecure person reacting to a difficult situation. Granted, this person has some power over you. That can be frightening, but what were you **really seeing**?

A good supervisor can deal with the poor performance of a supervisee in a way that will help the employee grow. Does yelling at a worker help that person grow? Does yelling communicate respect for another human being? Or does it indicate that the person yelling has something else going on? It is very important that you are able to see what is going on in a conversation so you don't just **react**. Getting caught up in the "heat of the moment" is **not** effective supervision. That brings us to the dimension of empathy.

## The Power Of Empathy

The dictionary says that **empathy** is "understanding so intimate that the feelings, thoughts and motives of one are readily comprehended by another." Quite a supervisory skill, don't you agree? Can empathy be learned? Everyone has some level of empathy; however, individuals can learn to develop their empathic skills.

Actually, empathy is a two dimensional skill. The first dimension is the complete *affective comprehension* of another person, which means understanding how the person is feeling and reacting emotionally. The second dimension is *effectively communicating your comprehension* of the person's situation back to him. Unfortunately, a course like this cannot give you empathy. You must take it upon yourself to practice perceiving other peoples' feelings, then reflecting those feelings back to them. You will not always be on target, but with practice you will become a more empathic supervisor. You will certainly learn more about yourself, which is where this discussion began.

## Respecting Individuals

Many times we get in our own way when talking with people. We build mental roadblocks to communication by maintaining prejudices such as, "What does she know about firefighting? She's just a rookie." The dimension of respect deals with how we sabotage the receipt of messages.

Everyone deserves **respect** simply because they are human. Our common life experiences bind us together, giving us a framework from which to learn and grow. When we let biases creep into our relationships with people, we limit our ability to truly communicate with the other person. Another way of looking at this dimension is to acknowledge that every person is worthy of *unconditional positive regard.* This means that we make an attempt to view each person in a positive light, even though we may not agree with them in all areas. The fact that this firefighter recently graduated from the academy shouldn't influence our opinion of him negatively. Or conversely, the fact that this person has been on line for 25 years shouldn't cause us to believe that he is a "dinosaur" simply counting the days to his retirement. The labels we place on each other limit what we can learn from them. If we decide that this person isn't worth listening to, how can we expect to have a supervisor/supervisee relationship?

## The Art Of Listening

Listening brings us to the last communication skill. **_Active listening_** is perhaps the most important communication skill you can develop. Once communication gets going, you need to learn how to keep it flowing. Active listening combines many of the dimensions we have discussed into a practical method of keeping the communication on track. The first part of active listening involves your nonverbal communication. **_Attending behaviors_** are those behaviors that communicate respectful attention to the other person. A slightly forward-leaning posture, eye contact, and appropriate touch can communicate your interest to the other person. The idea is to physically or nonverbally communicate to the other person that you are "there" with them. Next comes the verbal part. The receiver actively focuses his attention on what the sender is saying. Keep your thoughts, reactions, and prejudices at bay. Focus on what the other person is communicating to you. This can be the words themselves, but also the sender's emotions and nonverbal messages. You, the receiver, should try to understand what the other person is experiencing. Then put the message into your own words. Reflect back the content and feelings of the sender to him in your own words.

This accomplishes two things. First, it lets the sender know that you are listening to them and understand what they are trying to express. Secondly, it is a way of checking out your perceptions. Be careful not to judge or criticize. Just reflect the sender's content and feelings. Once you have completed a couple cycles of this reflection, the sender feels secure with you. Then you can start to add open ended questions to your responses. This will open new areas to the sender, allowing for deeper exploration. By using active listening techniques you demonstrate your interest and concern for the message sender.

*The idea is to physically or nonverbally communicate to the other person that you are "there" with them.*

## PERSONAL SIZE-UP

It is advisable for each of us to conduct a personal size-up. This involves looking inward to recognize one's strengths, weaknesses, attitudes, faults, and capabilities. Each of us should ask ourselves the following questions:

1. Who and what am I?
2. As a fire officer, I am (or will be) entrusted with the responsibility of leading in an environment of maximum stress and danger. Am I really qualified?
3. Have I acquired the technical knowledge and expertise necessary to take on this awesome responsibility?
4. Have I made a determined effort to learn the human side of supervision and leadership so that my subordinates and I will perform as a highly integrated team?
5. Have I studied and learned techniques which will enable me to motivate my subordinates so they can achieve their individual potentials?
6. Have I trained those I am in charge of in such a way that they are "we" oriented rather than "me" oriented?
7. Do I treat others fairly, but not all the same? Do I realize that each person has different and unique aspirations, abilities, idiosyncrasies, and drives?
8. Have I shown a willingness to share hardships with my people?
9. Do I see to the comfort and well-being of others before taking care of my own needs?
10. Have I accepted that the goals of my department take priority over my goals or those of my unit in particular incidents? Do I strive for goal convergence?
11. Can I issue distasteful, but necessary orders without alluding negatively to "higher-ups?"
12. Am I on an "ego trip," having become an ugly officer simply for self- gratification or to wear a brass-adorned uniform?
13. Am I an "officer," with all that the term entails? Do I look and act like an officer?
14. Can I show compassion and respect for those who attempt, but fail?
15. Am I aggressive without being foolhardy?
16. Have I kept abreast of changes in the ever-changing fire protection field or am I locked into the status quo?
17. Am I a suitable representative of my unit and my department?
18. Do I set a positive example?
19. Do I constantly strive to improve my performance and that of my unit?
20. Do I realize that having a sense of humor does not mean the ability to laugh, but the ability to laugh at myself when appropriate?
21. Can I sense the respect my people should have for me?
22. Am I physically fit, having strength, speed, and endurance?
23. Does our unit have a "can do" attitude? If not, why not?
24. Will my people be better firefighters for having served with me?
25. In summmary, am I a true leader?

After we individually work through these questions, we are confronted with the possibility that we need to modify our behavior and/or attitudes. How is this done? Where do we start? I suggest that we opt for education.

## What Knowledge Do I Need?

Too many of us suffer from a form of myopia or "tunnel vision," having depth but little breadth. Instead of being one dimensional people whose thoughts span from "A" for Apparatus through "H" for Hurst tool to "Z" for Ziamatic tools, we need those with greater "peripheral vision." We need "people for all seasons," who not only know that centrifugal pumps are not self-priming, but who also are familiar with such things as Abraham Maslow's theory of hierarchy of human needs and the philosophy of assessment centers.

In short, we in the fire service desperately require leaders who not only possess technical know-how, but who are also knowledgeable about all skills utilized during the major portion of their on-duty hours. Bear in mind that less than five percent of a firefighter's duty time is devoted to actual fire operations, with the balance taken up with activities requiring a type of supervision far different from that on the fireground.

I suggest each of us begin by making a commitment to self-improvement, followed by embarking on a program of study and training. On-the-job or trial-and-error learning to supervise is the worst type of learning. The result will be complaints of, "Yeah, his errors and our trials."

Many excellent books on leadership and supervision are available in your local libraries. Courses are taught in colleges and universities in personnel management and human relations. There are seminars conducted by various groups throughout the country that you may attend. Make your commitment and begin your journey!

(Written by Joseph Galvin, Suffolk County Community College, New York.)

# POST-PROGRAM ACTIVITY 1

*Instructions:* Circle the number that best describes how you would rank yourself in each of the following categories:

|  | VERY LOW |  |  |  |  |  |  |  | VERY HIGH |  |  |
|---|---|---|---|---|---|---|---|---|---|---|---|
| 1. Self-understanding | 0 | 1 | 2 | 3 | 4 | 5 | 6 | 7 | 8 | 9 | 10 |
| 2. Self-esteem | 0 | 1 | 2 | 3 | 4 | 5 | 6 | 7 | 8 | 9 | 10 |
| 3. Courage to fail | 0 | 1 | 2 | 3 | 4 | 5 | 6 | 7 | 8 | 9 | 10 |
| 4. Openness | 0 | 1 | 2 | 3 | 4 | 5 | 6 | 7 | 8 | 9 | 10 |
| 5. Peace of mind | 0 | 1 | 2 | 3 | 4 | 5 | 6 | 7 | 8 | 9 | 10 |
| 6. Tendency to trust | 0 | 1 | 2 | 3 | 4 | 5 | 6 | 7 | 8 | 9 | 10 |
| 7. Physical energy | 0 | 1 | 2 | 3 | 4 | 5 | 6 | 7 | 8 | 9 | 10 |
| 8. Versatility | 0 | 1 | 2 | 3 | 4 | 5 | 6 | 7 | 8 | 9 | 10 |
| 9. Creativity | 0 | 1 | 2 | 3 | 4 | 5 | 6 | 7 | 8 | 9 | 10 |
| 10. Flexibility | 0 | 1 | 2 | 3 | 4 | 5 | 6 | 7 | 8 | 9 | 10 |
| 11. Expressing your thoughts | 0 | 1 | 2 | 3 | 4 | 5 | 6 | 7 | 8 | 9 | 10 |
| 12. Listens to others | 0 | 1 | 2 | 3 | 4 | 5 | 6 | 7 | 8 | 9 | 10 |
| 13. Reaction to criticism | 0 | 1 | 2 | 3 | 4 | 5 | 6 | 7 | 8 | 9 | 10 |
| 14. Tolerance of difference | 0 | 1 | 2 | 3 | 4 | 5 | 6 | 7 | 8 | 9 | 10 |
| 15. Learning new skills | 0 | 1 | 2 | 3 | 4 | 5 | 6 | 7 | 8 | 9 | 10 |
| 16. Independence | 0 | 1 | 2 | 3 | 4 | 5 | 6 | 7 | 8 | 9 | 10 |
| 17. Vision of the future | 0 | 1 | 2 | 3 | 4 | 5 | 6 | 7 | 8 | 9 | 10 |
| 18. Goal setting | 0 | 1 | 2 | 3 | 4 | 5 | 6 | 7 | 8 | 9 | 10 |

|                          | **VERY LOW** | | | | | | | | | | **VERY HIGH** |
| --- | --- | --- | --- | --- | --- | --- | --- | --- | --- | --- | --- |
| 19. Handling hostility    | 0 | 1 | 2 | 3 | 4 | 5 | 6 | 7 | 8 | 9 | 10 |
| 20. Expressing opinions   | 0 | 1 | 2 | 3 | 4 | 5 | 6 | 7 | 8 | 9 | 10 |
| 21. Resisting pressure    | 0 | 1 | 2 | 3 | 4 | 5 | 6 | 7 | 8 | 9 | 10 |
| 22. Receiving affection   | 0 | 1 | 2 | 3 | 4 | 5 | 6 | 7 | 8 | 9 | 10 |
| 23. Giving support        | 0 | 1 | 2 | 3 | 4 | 5 | 6 | 7 | 8 | 9 | 10 |
| 24. Level of aspiration   | 0 | 1 | 2 | 3 | 4 | 5 | 6 | 7 | 8 | 9 | 10 |
| 25. Chances of promotion  | 0 | 1 | 2 | 3 | 4 | 5 | 6 | 7 | 8 | 9 | 10 |

In the following spaces, write the three (3) behaviors from the above list that you would most like to change. Briefly describe the actions you could take to change each of these areas.

1) Behavior:_______________________________________________

_______________________________________________

2) Behavior:_______________________________________________

_______________________________________________

3) Behavior:_______________________________________________

_______________________________________________

# POST-PROGRAM ACTIVITY 2

## SCENARIOS - MODULE I

### Scenario 1:

**Setting**: Scene takes place in the dayroom. The company officer is working at a table and is approached by a firefighter who wants to talk about personal problems.

**Firefighter** (sits down at the table where the company officer is working, lets out an exasperated sigh): That jerk really makes me mad!

**Company Officer** (puts pen down, looks directly at firefighter): What makes you mad, Bob?

**Firefighter**: Oh, Jack — he told me I didn't know how to rig the ropes on that last rescue.

**Company Officer** (leans slightly forward, intently): You felt he was unjustly criticizing you, then?

**Firefighter** (adamantly): Yeah! The system worked great. But he's always putting me down.

**Company Officer** (swings chair closer to firefighter): Always? Has he criticized you before?

**Firefighter** (angrily): Sure has. Just last shift he told me I wasn't mopping the right way! Can you imagine that? Mopping!

**Company Officer**: Sounds like you think Jack has it in for you. (phone rings, both men glance at phone, company officer pointedly ignores the phone) How can I help?

---

Evaluate this scenario and describe the positive and/or negative content. If you were the company officer, would you have done anything differently?

_______________________________________________________________

_______________________________________________________________

_______________________________________________________________

**Scenario 2:**

**Firefighter** (sits down at the table where the company officer is working, lets out an exasperated sigh): That jerk really makes me mad!

**Company Officer** (irritably glances up briefly from work) Huh? (looks back down at papers) What's that?

**Firefighter**: Oh, that clown Jack. He's on my nerves.

**Company Officer** (glances at watch while saying): Jack? Jack, who?

**Firefighter**: Jack Dobson, from E-1.

**Company Officer** (shuffles paperwork): Oh, yeah. Short guy, dark hair — seems like he knows his stuff.

**Firefighter** (angrily): Well, he certainly doesn't know everything!

**Company Officer** (keeps eyes on work in front of him, says in offhand manner): No, I guess not.

**Firefighter**: He complained about how I rigged the ropes on that last rescue.

**Company Officer** (still not paying much attention to firefighter): Well, he has taught the rope rescue course before. (phone rings, company officer reaches over to answer it and says) Maybe he was right.

**Firefighter** (pushes chair away from table, gets up and starts walking out of room saying as he goes): Maybe so, but it worked fine. I think he's just a jerk.

---

Evaluate this scenario and describe the positive and/or negative content. If you were the company officer, would you have done anything differently?

_______________________________________________________________

_______________________________________________________________

_______________________________________________________________

**Scenario 3:**

**Setting**: Scene takes place in the office of the battalion chief. The new company officer is coming to the battalion chief for advice.

**Battalion Chief** (smiles as company officer sits down): Hi, Bill. How're things going?

**Company Officer**: Well, I don't really know what to do about Johnson.

**Battalion Chief**: You have some sort of problem with Johnson?

**Company Officer** (hesitantly): Yeah, I do. Or maybe it's with me.

**Battalion Chief** (concerned tone of voice, but not pressuring): So, you're confused as to who really has the problem. What's been happening?

**Company Officer**: Well, he hasn't been his usual self lately. He's doing half-hearted work around the station. He seems to resist my direction.

**Battalion Chief**: And you're wondering whether his slacking off and resistance is because of you?

**Company Officer**: Right! I feel like I'm supposed to motivate him, but I just can't find the right buttons to push.

**Battalion Chief** (sympathetically): You're really frustrated because of the way he's been acting lately.

**Company Officer**: Yeah.

**Battalion Chief**: And you're blaming yourself because you think you're supposed to be able to motivate him.

**Company Officer** (says with relief): Exactly! Sometimes I wonder if I should ever have taken this promotion.

**Battalion Chief**: You're feeling insecure about being a new C.O. because of the way Johnson's been doing his work. (smiles) Hey, remember the old saying, "You can lead a horse to water but you can't make him drink?"

**Company Officer**: Yeah. So, there's nothing I can do about this?

Evaluate the preceding scenario and describe the positive and/or negative content. If you were the company officer, would you have done anything differently?

_______________________________________________

_______________________________________________

_______________________________________________

**Scenario 4:**

**Setting**: Scene takes place in the office of the battalion chief. The new company officer is coming to the battalion chief for advice.

**Battalion Chief** (smiles as company officer sits down): Hi, Bill. How're things going?

**Company Officer** (frustrated): I don't know what to do about Johnson.

**Battalion Chief**: Johnson? (laughs) He's no problem.

**Company Officer** (hesitantly): Well, he's sure a problem for me.

**Battalion Chief** (leans back in chair, puts hands behind head): You know, I used to work with him about ten years ago.

**Company Officer**: Really? I didn't know that. Was he always so half-hearted about his work?

**Battalion Chief**: Him? Nah...In fact, he usually worked real hard. (says smugly) You just have to know how to handle him.

**Company Officer** (worriedly): I guess that's what's wrong. I just can't seem to get him interested in anything.

**Battalion Chief**: Have you tried ordering him to do what you want?  You are his supervisor, you know.

**Company Officer**: I know, I just think there should be a better way. I don't want to force him to work. I want him to do it on his own.

**Battalion Chief** (forcefully): Sometimes you have to order a guy to do what you want. That's always worked for me.

Evaluate this scenario and describe the positive and/or negative content. If you were the company officer, would you have done anything differently?

_______________________________________________________________________

_______________________________________________________________________

_______________________________________________________________________

# REFERENCES — MODULE I

Kiersey, David & Bates, Marilyn. *Please Understand Me: Character and Temperament Types.* Del Mar, Calif.: Gnosology Books Ltd., 1984.

Lawrence, Gordon. *People Types & Tiger Stripes: A Practical Guide to Learning Styles.* 2nd ed., Consulting Psychologists Press, 1982.

Myers, Isabel Briggs. *Gifts Differing.* Consulting Psychologists Press, 1970.

# MODULE II

## LEAD, FOLLOW, OR GET OUT OF THE WAY!

**"When the best leaders' work is done,
the people say, 'We did it ourselves!'"
— Lao-Tzu**

# PRE-PROGRAM ACTIVITY 1

## STYLE QUESTIONNAIRE

*Instructions:* The following items describe aspects of leadership behavior. Respond to each according to the way you would be most likely to act if you were the leader of a work group. Circle whether you would be likely to behave this way:

Always (**A**), Frequently (**F**), Occasionally (**O**), Seldom (**S**), Never (**N**)

**As the leader of a work group, I would** ...

1. act as a spokesperson.    A  F  O  S  N

2. allow company members complete freedom of action.    A  F  O  S  N

3. encourage the use of uniform procedures.    A  F  O  S  N

4. permit members to use their own judgment in solving problems.    A  F  O  S  N

5. constantly supervise members for greater effort.    A  F  O  S  N

6. let members work as they think best.    A  F  O  S  N

7. keep work moving at a rapid pace.    A  F  O  S  N

8. turn members loose on a job, and let them go at it.    A  F  O  S  N

9. settle conflicts when they occur.    A  F  O  S  N

10. be reluctant to allow members freedom of action.    A  F  O  S  N

11. decide what to do and how to do it.    A  F  O  S  N

12. push for increased production.    A  F  O  S  N

13. assign members to particular tasks.    A  F  O  S  N

14. be willing to make changes.      A  F  O  S  N

15. schedule the work to be done.      A  F  O  S  N

16. refuse to explain my actions.      A  F  O  S  N

17. persuade others my ideas are to
their advantage.      A  F  O  S  N

18. permit the group to set its own pace.      A  F  O  S  N

# MODULE II NOTES

**This page intentionally left blank.**

# MODULE II

# LEAD, FOLLOW OR GET OUT OF THE WAY!

## OBJECTIVES

To provide the viewer and participant with:

- a basic understanding of the recognized styles of leadership
- an overview of the advantages and disadvantages of the different styles of leadership
- an assessment of their leadership and management style.

## OUTLINE OF KEY POINTS

**Expectations** — When an individual becomes an officer their relationships with superiors and subordinates will be different. An officer's ability to exercise leadership has a bearing on both personal and professional satisfaction.

**Between Superior and Subordinates** — Subordinates have a right to expect that their officers are capable of providing leadership and management under both normal and stressful conditions. Superiors have a right to expect their subordinate officers will carry out programs and decisions with the same degree of involvement that is the organizational norm.

**Between Peers** — Officers have the right to depend upon each other

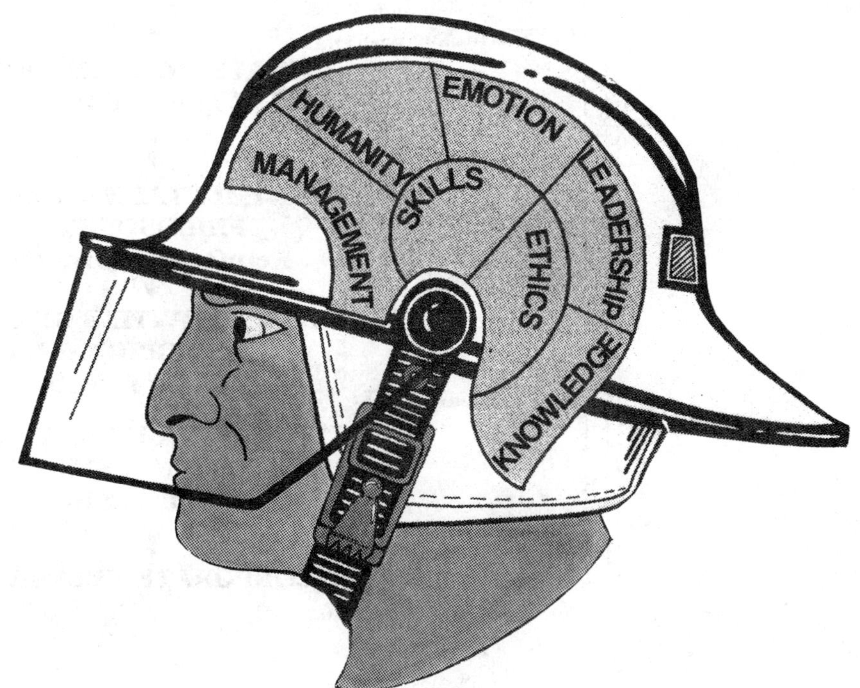


**Figure 2-1  Officer Development Skull**

to coordinate, cooperate, and communicate in the interest of themselves and the organization as a whole.

**Cultural Environment** — That unique combination of style, values, activities, and events that result in an organization's pervasive way of operating. The cultural environment is a constantly evolving process that affects productivity, conflict, stress, goal achievement, and organizational image. Corporate culture reflects the way things in an organization are allowed to be. *Negative corporate culture* means that there are no established value systems or direction. Each officer does whatever they want. Corporate culture can get better or worse as a result of management on a daily basis.

**Organizational Values** — That unique combination of perspectives that places value on a specific orientation to people, ideas, or tasks. The value system is what determines the relative merit of all activities in the organization. Each organization should have a set of values that everyone involved agrees upon. It should be an internal standard that anything done which goes against the set values is unacceptable. If your organization does not have an agreed upon set of values, then the values you operate by are simply a combination of everyone's unstated values. These values may be in conflict with one another, which can cause conflict and lack of direction for the organization.

**Roles and Activities of Fire Officers** — The fire officer is a pivotal point between the individual and the organization. Roles assumed by officers vary with local conditions.

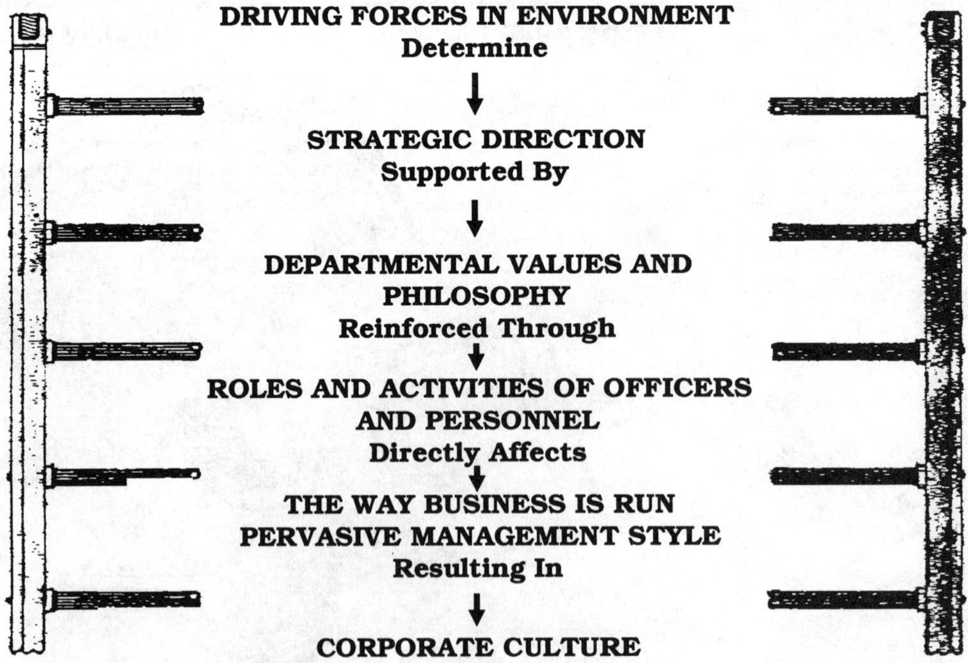


**Figure 2-2   Assessment of Cultural Change Needs**

Program activities determine workload. Individual orientation and style will determine the level of effectiveness in the organization.

**Organizational Environment** — The orientation used by leadership and management to accomplish the organization's activities. Individuals and organizations base much of their everyday interaction on one of the four orientations:

- **Task Orientation** — Focus is on the job, with little concern for its effect on individuals in the organization. Only concerned with completing the job, not what is being learned by the people involved.
- **People Orientation** — Focusing on the personal needs or emotional condition of individuals instead of the tasks to be performed. The focus can become one of making sure everyone is entertained and having a good time.
- **Goal Orientation** — Focusing on the maximum use of individual skills and abilities in achieving the objectives of the organization.
- **No Orientation** — Avoid the responsibility of accomplishing anything. This orientation disavows both the needs of the individual and the organization.

**The Difference Between Leaders and Managers** — Leaders do the right thing, managers do things right.

## McGregor's Theories and Other Leadership Styles

**Theory X** — The premise that people have to be closely supervised in order to get the job done.

**Theory Y** — The premise that if you leave people alone, they will do what is expected of them.

**Theory Z** — The premise that if you allow participation in goal setting, people will achieve.

**Theory W** — "Wishful Thinking Management." The premise that good intentions are all that is needed to provide direction in an organization.

**Theory U** — "Undesirable Management." The ineffective use of any style of managing an organization.

**Theory V** — "Visionary Management." The effective use of any style of managing an organization. This style places a high emphasis on team effectiveness and contentment, which leads to an effective team.

**Styles of Leadership** — Styles of leadership often develop out of how you expect others to respond to what you ask them to do. Do you trust your team to complete the task in the correct way? If so, then you will develop a different leadership style than if you do not trust the team. Some leadership styles are:

- **Delegating** — Deciding what has to be done and authorizing others to act upon the decisions.
- **Participating** — Allowing individuals to help set goals and act upon them.
- **Selling** — Deciding what has to be done and convincing others to support your ideas.
- **Telling** — Deciding upon what has to be done and ordering people to act upon it.

# WHAT'S YOUR LEADERSHIP STYLE?



A few years ago, there was considerable attention paid to a concept called Theory X and Theory Y. These two theories focused on two extremes of man's opinion of his fellow man. Theory Y saw people in a favorable light, ready to do the job with a minimum of supervision. Theory X saw people in an unfavorable light, ready to goof off at the slightest provocation, requiring constant supervision.

I imagine a fair percentage of the fire officers who studied these theories got caught up in the idea that everyone was either closer to Theory X or closer to Theory Y. The terms used for the two extremes were **dictator and democrat.** Considerable debate rose up among management students as to where they were on the scale. Of course, the primary focus for most people was to prove that they were in the middle, thinking this would imply fairness and assertiveness.

Nothing could be further from the truth. Besides, who cares where you are on a scale? Only two things really count about your leadership style is:

1. Are you comfortable with your leadership style?
2. Does your leadership style get the results you want?

These two questions become very important to the modern fire officer. If you can answer no to either question, then you have the potential for personal distress.

## Be Honest With Yourself

Let's talk about you for a moment. Think about the two questions asked above. What are your answers? If you answered "yes" to question one, then you are probably comfortable with yourself as a person and have a good self-image. That's good.

If the answer to question two is "yes" also, then I wonder why you would even need to keep reading. If your style works it's because you have found it successful under a variety of circumstances and conditions. You are probably already adept at the subtle changes required when situations move from low risk to high risk or vice versa. This doesn't mean you can't improve on your execution of leadership in a given environment. We can all improve. However, two "yeses" mean you're feeling O.K. about who you are and what you are doing.

Now, let's say the answer to number one is "no." You're not comfortable with your leadership style. Perhaps you feel like a square peg in a round hole. Don't feel alone. A lot of people have been trapped into "adopting" a leadership style that simply is not them. This occurs when you try to conform to a role model advocated in some leadership classes. It can also come about from trying to conform to someone else's perception of what you should be.

Let's look at a few examples. One officer I knew was basically a very humanistic person. His personality was structured around working closely with people and soliciting help. He, in turn, was always available to give support to anyone who asked for it. He was a successful captain. The troops saw him as a real "team man," a good leader.

Promotion time came and he made battalion chief. Someone told him his "style" was too loose! He was perceived as being too "laissez-faire." He was told to "toughen up." Boy, did he ever! Overcompensating for his natural style, he became a strict disciplinarian. The result was that he was not happy with himself and his subordinates were unhappy with him. He went from one confrontation to another as an unhappy person.

## Lead As You Are

Even though the above example is extreme, it is worthy of consideration. Your basic style of leadership is your perception of yourself. The more you know about yourself, the more you gain in leadership attributes. Let's repeat that statement so you'll get the full impact of it. **The more you know about yourself, the more you will gain in leadership attributes.**

What we are saying is that too many officers study "leadership," but they don't examine themselves closely enough to maximize their own skills in leadership. Every fire officer has an inventory of hidden skills they have been forming since childhood. These skills were exercised in kindergarten, on the baseball fields, in high school classes, and so forth. These hidden skills and the person's basic philosophy are extremely important in developing a sense of self-awareness.

What I am suggesting here is that a person has to "know himself" before he can utilize any of the so-called leadership attributes. Quite frankly, there is not "one" leadership style that is **the** style of success — not in the fire service, not in the business world, not in the military.

For example, can you see the distinct differences in the leadership style of Dwight Eisenhower, George Patton, and Douglas McArthur? All three men were successful leaders. All three had a great deal of confidence in their styles. They did not adopt a certain style as a leader. They were who they were.

More specifically, if your answer to question number one ("Are you comfortable with your leadership style?") is "no," then look first to yourself. Ask yourself, What am I comfortable with?" There are hundreds of evaluation instruments that are used by professional school counselors to assist you in knowing yourself better. Locate someone nearby who can aid in developing a personality profile to see what kind of person you really are. Examples of testing instruments commonly used are:

- FIRO-B
- MMPI (Minnesota Multi-Phasic Personality Inventory)
- LIFO (Life Orientation Analysis)
- LEAD (Leadership Effectiveness and Adaptability Description)

The list goes on and on. Any college counselor can help you locate someone who is qualified in administering these instruments.

By using these devices, you can often see what your comfort problem is. For example, your FIRO-B score could show that you do not like to have someone exercise strict controls over your life, yet you like to exercise control over others. Then if you are in a situation where most of the tasks you are given are mandated by a superior, you may adopt an aggressive style that keeps you in distress most of the time. By the way, this is not an uncommon problem for staff officers.

The bottom line of this discussion is that the ***comfort factor*** as a leader means possessing a sense of being comfortable with your basic personality and knowing how it effects your leadership style.

I can almost hear the arguments now. I'll bet there are those of you who claim to be one person on the job and another person off the job. It might be true, but I'll bet it causes conflict. Sometimes it causes confusion on the part of subordinates and superiors alike. They don't know how you'll react unless they can see what kind of attire you are wearing! Badge on — leadership on. Badge off — nice guy, or whatever. To increase your comfort as a leader, know yourself and be satisfied with who you are!

## Are you Getting Results?

Now to question number two, "Does your leadership style get the results you want?" If you answered this question "no," what can you do? There are two options:

1. change yourself to conform to a new style and hope it works
2. stay yourself and study your subordinates to see why nothing is happening

Option number one has obvious pitfalls. Refer back to the discussion on question number one. You cannot change who you are to fit a leadership style. You must find a leadership style that fits who you are and work on changing the negative parts of your personality.

Option number two infers that to be an effective leader, you must understand the expectations of the people you are trying to lead. In short, this means your leadership style should match the leadership needs of your group. The group's perception and acceptance of your leadership style will effect interactions with you.

Another example might help here. A fire department that had operated under an open, flexible management style changed chief officers. The previous chief delegated authority, solicited input, and developed decisive officers. Subordinates felt they were a team and the chief was the coach. It was a successful combination.

When the chief moved on, a staff officer became chief. He did not understand the system even though he had been promoted from within the organization. He perceived the system as too loose. He increased demands, pulled back authority, and was not open to constructive criticism. He saw himself as a dynamic leader, the troops saw him as an interference. **The result — poor performance.**

Even though the previous chief used a laid-back leadership style, more work was accomplished. What should be done? The new chief could have been a dynamic leader without destroying his relationship with the troops. He doesn't need to change himself, he needs to closely evaluate the profile of the group he is leading. He needs to understand their collective strengths and weaknesses and mix his needs with theirs.

*A new officer needs to understand the collective strengths and weaknesses of his/her subordinates and mix and match his needs with theirs.*

## Pulling The Team Together

In developing a group profile, team builders can use the testing instruments referred to in our discussion of question number one (page 45). The leader can analyze what the group expects of him. Without changing his style, he can begin to meet the needs of the group by avoiding unnecessary conflict.

This may sound deceitful or opportunistic, but it isn't. You have probably heard the saying, "do unto others as you would have them do unto you." That may not hold true for a leader. To motivate effectively, a leader should do unto others what they want done to them. Knowing what makes people tick can be an invaluable tool in structuring your leadership role.

The most successful team managers I know of are ones who have developed an effective leadership style. They do not try to make anyone into their image. They have the ability to recognize the style of their subordinates and encourage them to use their strengths to the maximum. They compliment the weaknesses of each style with the strengths of another. If the team is as sold on a concept as the leader, tasks will be accomplished. A natural leader intuitively encourages and motivates his team. He knows himself and his followers. However, some of us have to work at developing an effective leadership style.

If you answered both questions "yes," then go out and help someone else get two "yeses." If you have one or two "no's," then begin studying your strengths and those of the troops. There are no bad leadership styles or bad followers. There are only uninformed leaders and mismanaged followers. To borrow a thought that was recognized long ago, if you want to be a leader, lead thyself!

*There are no bad leadership styles or bad followers. There are only uninformed leaders and mismanaged followers.*

## THEORY V

We all have a tendency to put labels on people. In the fire service, we talk about a person being a particular type of leader or manager. Unfortunately, labels often fail to stick because people change. Also, theory and reality don't always jive.

In his textbook, *The Human Side of Enterprise*, Douglas McGregor describes basic theories of leadership style. He labeled them Theory X and Theory Y. **Theory X** was based on the premise that people are not good at supervising themselves and have to be told what to do. In its most simplified version, Theory X managers were seen to be **autocratic**, or as having the absolute power in the organization. Many read this to mean dictatorial.

**Theory Y** was based on the premise that everybody wants to do the right thing. Subordinates will always do what has to be done if given the chance. Theory Y managers were seen to be **democratic.** Some interpreted that to mean laissez-faire, or a noninterfering style.

We, in the fire service, have a problem with McGregor's dichotomy because both Theory X and Theory Y styles are needed if we are to succeed. The fireground or scene of a major medical emergency is not the place to practice decision making by committee. Participatory management doesn't function under high stress. Fireground commanders must be dictatorial. At other times, the rule of the majority is not only traditional, but essential in making the system in the fire station work. Try dictating what will be served at a company's evening meal, or controlling what will be eaten for the next ten shifts.

To complicate things further, someone came along and added the letter Z to this alphabet soup. If the fire service thought it had problems with Theory X and Theory Y, it was in for big trouble with Theory Z. Imported from Japan, **Theory Z** was advocated as a remedy for production problems. Theory Z holds that if everyone is allowed to participate in goal setting, people will achieve more.

### What Do They All Mean?

To put all these management styles into perspective we must discuss the leadership and management model proposed by McGregor. The model he used to illustrate this concept was a graph, where Theory X was placed on one axis and Theory Y on the other. The inference is that a person could adopt traits of each leadership style and end up being placed on the graph at the halfway point.

McGregor was the first psychologist to identify the fact that styles of management were dependent upon a manager's basic philosophy about people. He identified what we might consider to be two extreme belief systems about people: Theory X and Theory Y.

However, there are points that lie between Theories X and Y that may help you identify characteristics in yourself or others. This understanding can significantly aid you in sharpening your management skills.

It is plain to see that a manager who sees people who are mostly Theory X types would have a tre-

> *There are points that lie between Theories X and Y that may help you identify characteristics in yourself or others.*

mendously different management style when compared with a Theory Y believing manager.

Let's briefly review some of the things a Theory X manager would say or do. Also look over some personality traits of a Theory X manager so they may be more easily spotted and modified when appropriate.

### What You May Hear From A Theory X Manager

1. Do it!!
2. Don't ask questions.
3. Do it my way.
4. This is the way we've always done it.
5. If you want anything done right — do it yourself.

### Personality Traits Of A Theory X Manager

1. Super critical.
2. Blunt.
3. Disregards suggestions or takes personal credit for the suggestions of others.
4. Defensive.
5. Narrow-minded.
6. A poor delegator.
7. Over involved in detail.
8. Negative attitude toward those who work for him/her.
9. Insecure — requires frequent encouragement.
10. Insensitive.
11. Suspicious.
12. Egotistical, at least outwardly.
13. Ruthless.
14. Communicates downward only with those who work for him/her.
15. Never asks — gives orders.
16. Would be likely to use performance reviews as a weapon — another vehicle with which to criticize. Could consider reviews to be a waste because of basic beliefs about people.
17. Would be a (10, 1) manager on the grid — Task Management Oriented.

*Beliefs about the basic nature of humans causes an officer to adopt a particular management style.*

## Get The Picture?

Whether you fully agree with all the factors outlined above is not as important as keeping in mind the major point — these beliefs about the basic nature of human beings lead to a particular management style.

Theory Y believers are described in almost exactly the opposite way as Theory X believers, so a listing of their traits is not necessary.

Theory Z managers believe that most people want to participate in goal-setting and in decision-making. If workers are encouraged and allowed to participate, and their ideas and suggestions are taken seriously, they will be motivated to achieve the goals of the organization, as well as their own personal and professional goals.

### Assumptions Held By Theory X Supervisors

1. The average human being has an inherent dislike of work and will avoid it if he can.
2. Because of this dislike of work, most people must be coerced, controlled, directed, and threatened with punishment to get them to put forth adequate effort toward the achievement of organizational objectives.
3. The average human being prefers to be directed, wishes to avoid responsiblity, has relatively little ambition, and wants security above all.

### Assumptions Held By Theory Y Supervisors

1. The expenditure of physical and mental effort in work is as natural as play or rest.
2. External control and the threat of punishment are not the only means of bringing about effort toward organizational objectives. People will exercise self-direction and self-control in the service of objectives to which they are committed.
3. The average human being learns, under proper conditions, not only to accept responsibility, but to seek it out.
4. The capacity to exercise a relatively high degree of imagination, ingenuity, and creativity in the solution of organizational problems is widely, not narrowly, distributed in the population. In other words, the ability to be a creative problem-solver exists in large numbers of individuals, at all levels within the organization.

### Assumptions Held By Theory Z Supervisors

1. No matter what a person likes or dislikes about their work, they will achieve more if they are given an opportunity to use their own strengths to the advantage of the organization.
2. The average human being loves to be on a winning team.
3. The capacity to succeed is omnipresent under even the most adverse of circumstances.

## The Leadership Grid

An *INC Magazine* article included a graph entitled "The Leadership Grid" that provides another explanation of leadership styles. On the Leadership Grid (Figure 2-3), the axis labeled "Concern for Task" is equivalent to Theory X and the axis labeled "Concern for People" is equivalent to Theory Y on the McGregor graph spoken of earlier. This grid says nothing about the style a person uses to accomplish the task. It merely suggests that some people are more people oriented, while others are more task oriented. The manner used to achieve that objective may or may not be marked by a particular management theory.

The leadership grid model uses numbers 1 through 10 on the two different axes. These two axes, as we mentioned earlier, are labeled Tasks and People. The inference of this model is that a person can be oriented toward either completing tasks (T) or meeting the needs of people (P). An individual can be 1 at the task (T) level and 10 at the people (P) level, or 10 at the T level and 1 at the P level, or 10T and 10P. The leader who adopts a style of being high task oriented, but has a low concern for people will find it difficult to get much of a response

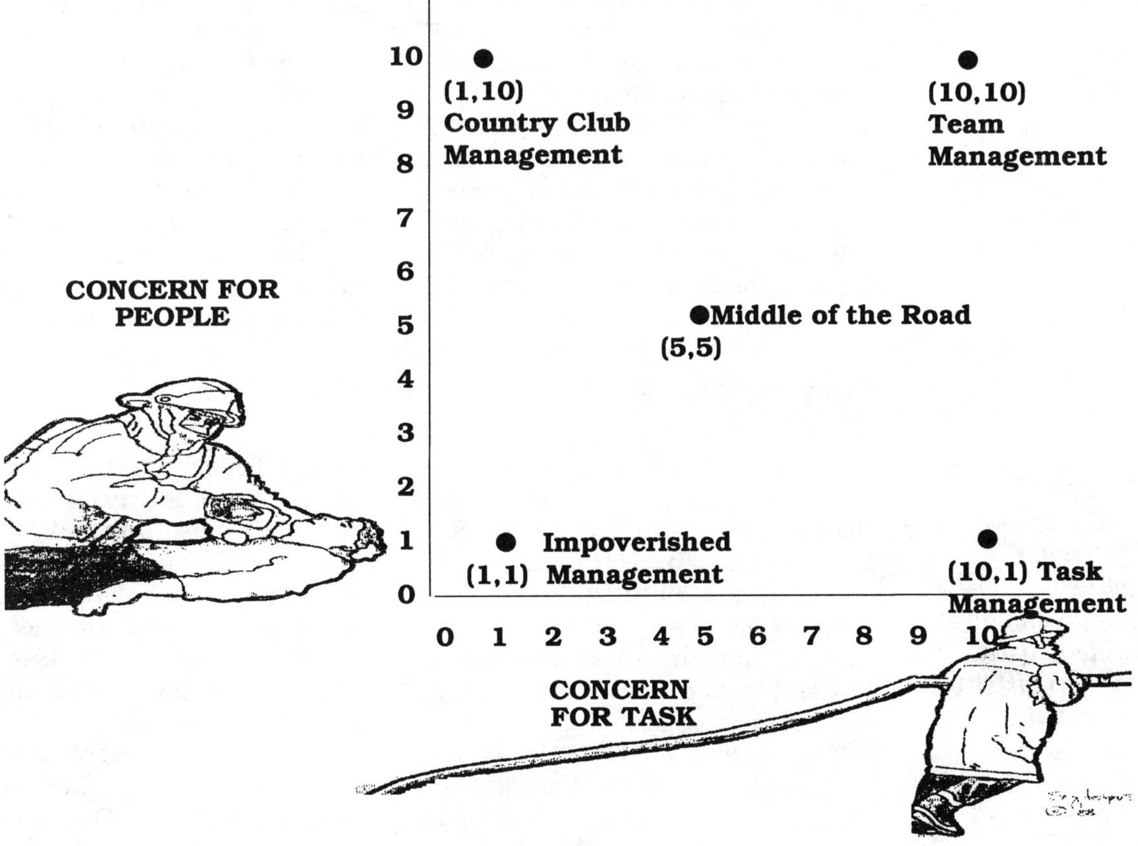


**Figure 2-3   Leadership Grid**

from the individuals he/she is leading. In the center is a group of individuals who try to please both sides by being 5-5.

Since Theory X is oriented toward tasks and Theory Y toward people, where does that leave Theory Z? According to the literature, Theory Z most likely fits into one of the opposite corners, since the Japanese point out that their country is the most productive and the most people oriented.

If the Leadership Grid and McGregor's model are overlayed, it is possible to see that there is a relationship between the two. When viewed from a management theory standpoint, the Leadership Grid can be an aid in explaining management "styles."

If we suppose that these two grids are overlayed, as previously discussed, we can see that a 1T-10P manager would tend to be democratic, or fall into Theory Y management style. A 1P-10T manager would tend to be task oriented and represent Theory X management style. This sounds simple enough.

What about 1T-1P managers? Is there a theory describing them? I suggest that there is, **Theory W.**

## Finding Your Style

We have Theories X and Y provided by McGregor, Theory Z from the Japanese, and Theory W coming out of apathy and indifference. In Figure 2-4 all of the leadership theories are combined to illustrate that a leader may use situational leadership and flow in and out of the different styles.

The bottom of the cube is the undesirable application of a leadership theory, or Theory U. The top of the cube is the desirable, productive utilization of a leadership the-

The W stands for Wishful Thinking. If a leader is low on their concern for both task and people it would follow that it will be only wishful thinking to anticipate that the organization will ever accomplish anything. Chances are that a 1T-1P organization may be able to muddle along for awhile, but sooner or later it will die of obsolescence.

While few fire administrators would admit to the label, Theory W does exist as a style of leadership/management in many organizations. In fact, the more the leadership denies the use of Theory W in daily activities, the more likely the theory is alive and well in the practices of the department.

Placement of an individual on one of these grids should not imply that this person will behave a particular way all of the time and in all circumstances. People rarely behave in only one way. Most of us modify our behavior in different situations. When a leader effectively uses different behaviors in response to different situations, we call it **situational leadership.** The problem with this label is that it can make the leader sound uncommitted or even manipulative.

ory, or Theory V. The four walls of the cube represent the four leadership styles discussed: W, X, Y, and Z. As the cube is rotated on it's four sides, you see that the outcome of using any of the four leadership styles may be either desirable or undesirable depending upon the circumstances.

**Theory U** is based on the premise that any style can be used in a negative fashion. Democracy can be taken to the level of anarchy and autocracy to the level of dictatorial

*It doesn't make any difference which management style is used. What matters is how it is applied and the results that are obtained.*

dominance. Theory U stands for Undesirable. Any leadership style can be made undesirable if the individual fails to realize its implications and ineffectiveness.

***Theory V*** stands for Visionary. Theory V is based on the premise that any management style can be used in a productive and useful manner, if the user has an adequate understanding of how to effectively use that particular leadership style. However, the user must have a sufficient grasp of the organizational situation and how continued use of that style will effect his/her subordinates and organization.

In other words, it doesn't matter which management style is being implemented. What matters is how the management style is applied and if the desired results are obtained.

Too many people fall into the trap of believing that one management style is more effective than all others. That is simply not true. Different people under different circumstances use different management styles both effectively and ineffectively.

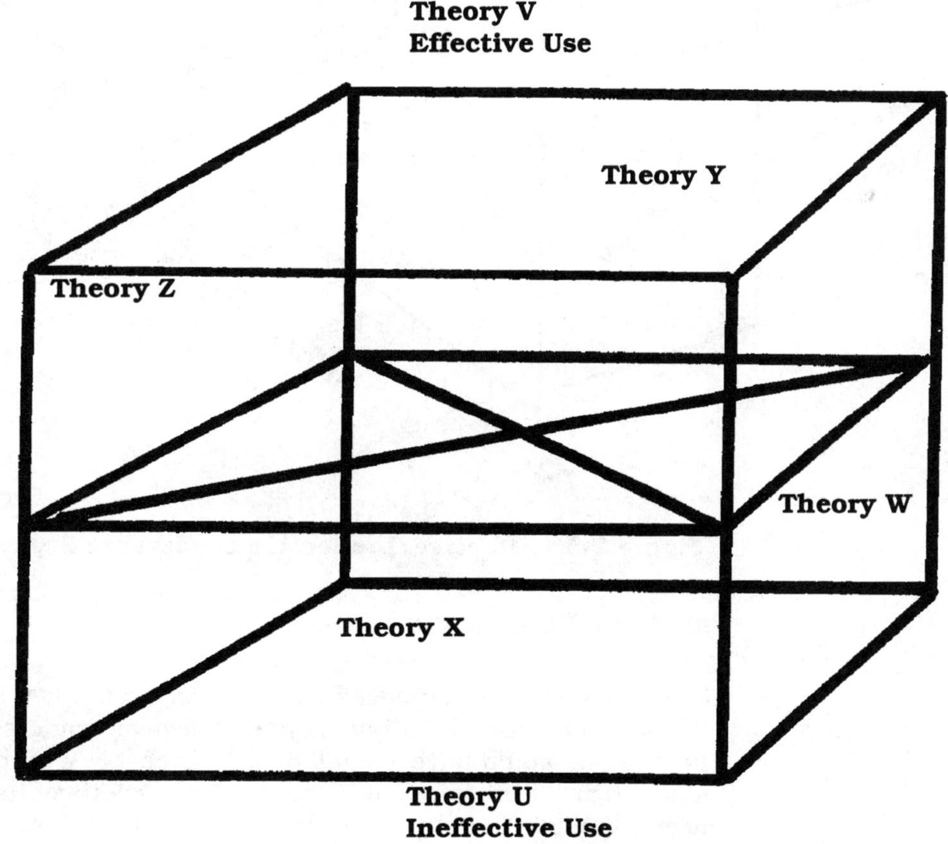


**Figure 2-4   Theory V**

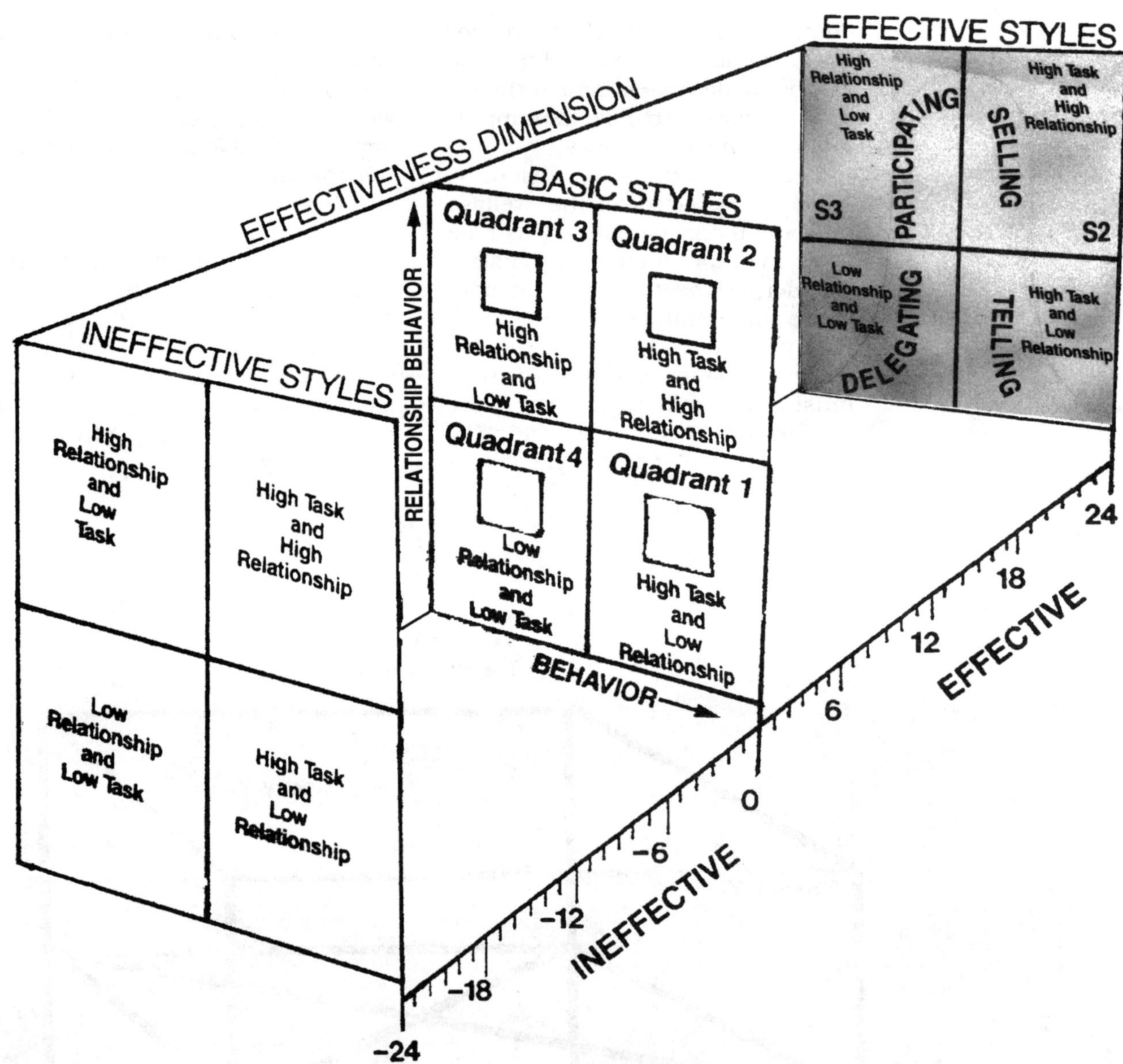


**Figure 2-5  Effective/Ineffective Leadership Styles**

## Making Your Style Work

If we consider a manager's effectiveness as being like a box, some individuals would have a small box since their effectiveness as a manager is limited. This concept is illustrated in Figure 2-5, Effective/Ineffective Leadership Styles[1]. The degree to which an individual is conscious of using or misusing the different leadership styles varies greatly.

---

1  Paul Hersey/Ken Blanchard, *Management of Organizational Behavior: Utilizing Human Resources*, 4/e, (c)1982, P. 163. Adapted by permission of Prentice-Hall, Inc., Englewood Cliffs, New Jersey.

If you study the back panel of Figure 2-5 carefully, you will see there are four effective leadership styles.

- When there is **low task** (the task is relatively unimportant) and **low relationship** (little interaction with the leader is required to accomplish the task), an effective leader would use **Delegation.**
- When there is low task and **high relationship** (much interaction with the leader is required to accomplish the task), the leadership must use **Participation.**
- Where there is **high task** (the task is important) and low relationship, you must **Tell** people what to do for you to be an effective leader.
- When there is high task and high relationship, you must **Sell** people on your ideas in order to be an effective leader.

Effective leadership styles are situational. There is no single best leadership style. Effective management styles tend to respond to whatever is happening in the organization at that time. For example, an effective fire officer would use a **Participative** management style when programs are being managed and a **Telling** style in an emergency situation.

It should be apparent that a person who uses a Theory V style of leadership and a strong sense of reality is able to mix the task/people elements with their leadership techniques to be an effective leader. It is also easy to visualize that a Theory U style is almost always based on ego and self-centered behavior regardless of the individual's orientation toward tasks or people.

People sometimes find it difficult to identify their leadership style. They feel uncomfortable being plotted on a graph or placed in a box that is based on assumptions provided by others. However, if we accept the existence of Theory V we can explain our choice of leadership style and behavior based on what we feel the future holds.

You may be a taskmaster, but you have a game plan in hand. You may be a humanist, but you are not relinquishing your goals and objectives because of the desires of others. An individual should be able to explain his/her leadership style, not in terms of Theory X or Theory Y, but in terms of how and why they have adopted a certain leadership style.

Granted, most people don't like to be pigeonholed, but Theory V allows us to be put in a box of our own construction. It can be as small or as large as we desire. We are the carpenters — it is our construction, our design. If it fails, we suffer the consequences. If it works, we get the credit.

*An individual should be able to explain his/her leadership style in terms of how and why they have adopted a certain leadership style.*

# LEADERSHIP IS A MATTER OF PERFORMANCE

I am going to ask you a history question. What do these leaders have in common?

a. Mohandas Ghandi
b. Adolph Hitler
c. George Patton
d. Dwight D. Eisenhower
e. John F. Kennedy
f. Ronald Reagan

In the case of the first two men, one was a man of peace, the other was a man of war. The second two men were both military leaders. One was flamboyant, the other was a study in moderation. The two presidents were probably as diverse as you can imagine. One was a Democrat, the other a Republican. One was liberal, the other conservative.

So, what do they share in common? The answer is that they all have a place in history because enough people believed in them to make them visible in the eyes of the world. They became leaders because they had characteristics that caused others to follow.

In the fire service we talk about leadership, but it is often placed in the context of management and/or authority. If we examine the list of names above, we might be hard-pressed to find a common thread regarding their personalities or characteristics. What is it that these individuals shared that allowed them to be selected by large numbers of people to serve in a leadership capacity?

When discussing this issue, many personality traits begin to emerge. Styles and types, such as being "likeable" or "perserverant," become apparent. Characteristics, such as intelligence and courage, emerge at the forefront. Often the characteristics we like in others are what we look for in relationships with people, but they are not necessarily a part of leadership. For example, could Adolph Hitler be construed as "likeable?" Perhaps to his closest friends, but to millions of people he was a threat to their lives. If we study leadership styles in terms of common personality traits, we will be disappointed. Leaders come in all sizes and shapes. In the words of John Gardner, "Leaders can be found scattered all over the place."

Is there one trait they share? The one characteristic they all display is they have the ability to perceive what other people want or need and give it to them. Gandhi stated it succinctly, "There go my people — I must hurry up and get in front of them, for after all, I am their leader!"

*The one characteristic this list of leaders display is that they have the ability to perceive what other people want or need and give it to them.*

## Who Are The Leaders?

In the context of public services, we often mistake authority with the ability to serve in a leadership capacity. Many individuals who emerge in the top jobs of fire departments have strong leadership characteristics. However, there are also those with good leadership abilities at the division, platoon, and company level. In fact, there are formal and informal leaders at every level of organizations. This is definitely an advantage in the fire service. Unfortunately, a lot of these leaders are not given the opportunity to develop. In some organizations it is often considered a threat to those

56

in authority if members exercise leadership.

All fire officers have to be leaders in one respect. Namely, they have the responsibility to take their organization and move it from where it presently is to where it eventually needs to be. Failure to exercise that leadership responsibility results in an organization becoming stagnant.

It pays for us to take a look at leadership traits as they exist vertically and horizontally in our organizational environment, for there is plenty of room for leadership in the fire service. Many people today perceive that those who hold the bureaucratic offices or high positions in the organization are the true leaders. However, this perception ignores the fact that true leadership has nothing to do with the position a person has in the organization. For example, sometimes leadership exercised by individuals who are not delegated power will result in the overturning of those in authority over them, as well as the power structure that resists their motives and/or goals.

So, what leadership traits are we looking for in our organization? Where are the leaders? Who are the leaders? What traits do they demonstrate that make them the potential source of energy for organizations?

## Leadership Means Moving Forward

There are some traits that all leaders share. Leaders are:
a. goal oriented
b. change oriented
c. future oriented
d. and possess these traits internally.

Simply stated, no one ever became a leader by standing up and stating, "Let's keep on doing exactly what we have been doing!"

A quick overview of what is going on in the fire service today indicates that we have a strong need for leadership to emerge at every level of the fire service. The fire officers of today have a responsibility to exercise leadership through activity. Obviously, it is anticipated that we will fulfill this capacity in handling emergency operations. More importantly, we must expect to exercise leadership in fire prevention, public education, public policy, and develop of a philosophy that will transcend the end of one century and the beginning of another.

This is an interesting time to be a fire officer! A century of progress has charted the growth of fire protection. Philosophical and technological changes are occurring in the fire service everyday. There is opportunity on many levels to exercise the leadership role.

Some officers will succeed in exercising leadership and others will fail. In the near future, all officers will have the opportunity to make a decision to move forward or back away.

Most of us will not have the opportunity of someone like Lee Iacocca, who was able to move from a hostile arena to the cover of *Time Magazine*. They **will**, however, make a difference on the local, regional, state, and national fire scene if they care to do so.

*No one ever became a leader by standing up and stating, "Let's keep on doing exactly what we have been doing!"*

## Ground Rules Of Leadership

There are some ground rules for becoming a leader in the fire service that have nothing to do with rank, the size of the fire department, race, or religion. The ground rules consist of four basic premises:

- you must believe in something more than yourself
- you must learn to be a good follower before you can ever exercise leadership
- you must focus on meaningful change instead of change for the sake of change,
- and you must be committed to long term gains rather than short term wins.

One interesting aspect of leadership is that it seldom occurs spontaneously. Generally speaking, an individual assumes a leadership role after he accumulates a considerable amount of experience in interacting with those he intends to lead. This is not unique to the fire service. Developing into a successful leader takes considerable time. As individuals emerge as potential leaders in various levels of the organization, the same rule applies — **patience is absolutely necessary if an individual wants to assume a leadership role**.

Leadership may also be compared to a "ship." Like a naval vessel, leadership is meaningless unless it is going somewhere. Leadership must be navigated, it must have a course plotted.

Later in this course we will discuss the concept of mission statements, goals and objectives, and action planning for the fire department. One of the most critical elements in exercising leadership is to utilize these techniques to give direction and structure to the individuals who have responded to that leadership call.

> *One of the most ciritical elements in exercising leadership is to give direction and structure to the individuals who have responded to that leadership.*

## Navigating To Success

Like a large naval vessel that sails the seas, a leader must have a crew to help in operations. The concepts of team building, interpersonal dynamics, and group behavior are not theories, they are a reality in leadership. Without a team or recognizing the strengths and weaknesses of that team, no leader can function for long. Without a sense of perspective of the value and contribution of all the members of the organization, leaders often find themselves isolated from their followers and subsequently lose the mantle of leadership.

No vessel can continue to pursue a course unless it has either wind in the sails or fuel in the bunker.

Resources are invaluable in exercising leadership. They can be obtained or denied from others who are exercising leadership. The leader of today is not an individual who wishes to take away from someone else, but rather reinforce and build upon the strengths of individuals with similar interests and goals. It is important to note that resources are transferable. That is to say time, money, manpower, and authority can be taken away from an individual and given to someone else if the leader does not exercise caution.

Periodically we need to look at the compass. No matter where you intend to go in your role as a leader, there are obstacles that cause you to deviate from your course. The hand on the tiller of leadership should utilize evaluation techniques, monitor progress, and solicit feedback on how things are going. In the fire service this means that leaders listen almost as much as they talk. Feedback can be positive or negative. You cannot afford to ignore potentially harmful situations or bask in the glory of another.

Among the most severe obstacles in handling a ship is changing weather and the effects of tides and currents. The counterpart in the fire service is the rise and fall of political or governmental influence and the trends in local government. In exercising leadership in the fire service, we need to keep a watchful eye on these conditions. Effective leadership includes knowing when to go with the trend and when to oppose it. Changes in political climate are often subtle and at other times stormy! It requires skill to stay on course.

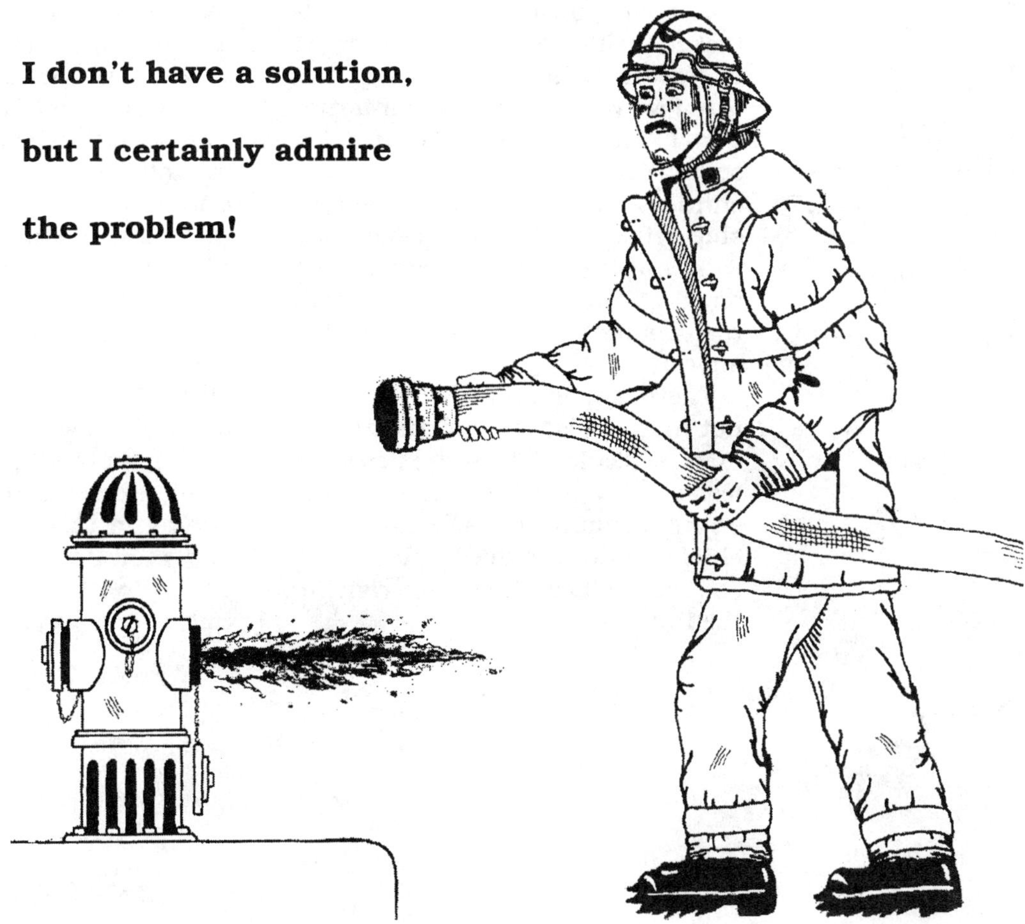


**Figure 2-6   I Don't Have the Solution**

## Will You Lead Or Follow?

At the beginning of this module, we talked about different leadership styles. In the fire service we have many types of leaders. The historical figures mentioned on page 56 can be viewed positively or negatively. There are good leaders and there are bad leaders. The exercising of leadership is totally independent of its morality or value system.

There are good fire service leaders and bad ones. Contrary to conventional wisdom, leadership does not always result in rewards. There are those who exercise leadership and have subsequently been attacked, criticized, and even forced to give up the leadership role.

A few years back, the International Society of Fire Service Instructors printed an article entitled, "The Paradoxes of Leadership." It stated, "The biggest of men with the biggest of ideas can be shot down by the smallest of men with the smallest of minds. Think big anyway."

There are literally hundreds, if not thousands, of opportunities to exercise leadership in the fire service. The fire chief can do it for the entire community. The battalion chief can do it within a platoon. A company officer can do it right in the firehouse.

There is opportunity to exercise leadership in civic groups and professional associations. There is no shortage of leadership opportunity. There **is** a shortage of effective leaders. ISFSI discussed some paradoxes that are often used as excuses for not becoming involved as a leader. One that spoke to me was, "If you do good, people will accuse you of being selfish or of having ulterior motives. That's true, but do good anyway."

As you have gone through this module one thought has probably gone through your mind — not everyone can be a leader or there won't be any followers. One thing that gives an organization great strength is not the identity of the leader, but the collective identity of the followers. A winning combination can evolve if you recognize what leadership is, respect it for what it can do, and support it in a positive fashion. It is fine to choose not to step out in front and lead. To choose to support those that do is also OK. Being part of a winning team is a reward in itself.

So remember the short, but succinct, expression of leadership — **Lead and I will follow you, or follow me, I will lead. But in any case, don't just sit there — get out of the way!**

*One thing that gives an organization great strength is not the identity of the leader, but the collective identity of the followers.*

# THE PARADOXES OF LEADERSHIP

People are illogical, unreasonable, and self centered. Love them anyway.

If you do good, people will accuse you of selfish, ulterior motives. Do good anyway.

If you are successful, you win false friends and true enemies. Succeed anyway.

The good you do today will be forgotten tomorrow. Do good any way.

Honesty and frankness make you vulnerable. Be honest and frank anyway.

The biggest men with the biggest ideas can be shot down by the smallest men with the smallest minds. Think big anyway.

People favor underdogs, but follow the top dogs. Fight for a few underdogs anyway.

What you spend years building may be destroyed overnight. Build anyway.

People really need help, but may attack you if you do help them. Help people anyway.

Give the world the best you have and you'll get kicked in the teeth. Give the world the best you have anyway.

# POST-PROGRAM ACTIVITY 1

### HOW TO SCORE YOUR STYLE QUESTIONNAIRE (PRE-PROGRAM ACTIVITY 1)

A) Circle the statements on your Style Questionnaire numbered 1, 3, 9, 10, 11, 15, 16, and 17.

B) Write a "1" in front of the above numbered items to which you responded S (seldom) or N (never).

C) Write a "1" in front of the items not circled to which you responded A (always) or F (frequently).

D) Circle the number "1's" you have written in front of the following statements: 2, 4, 5, 6, 8, 10, 14, 16, and 18.

E) Count the circled "1's." This is your score for concern for people.

F) Count the uncircled "1's." This is your score for concern for the task.

The average manager scores 3 on task orientation and 5 on people orientation. If you score more than 3 T's and fewer than 3 P's, you have a strong affinity for getting the job done, you work better alone, and you're good at staff and technical work, but poor at managing people. The reverse is true if you score fewer than 3 T's and more than 3 P's.

As a rule, managers with high T scores and low P scores don't work well together. However, the addition to this group of just one manager with a high P tally results in a highly effective decision making team.

# POST-PROGRAM ACTIVITY 2

**SCENARIOS - MODULE II**

**Scenario 1:**

**Setting:** Scene takes place in cab of engine upon arrival at a car wreck. A company officer and two firefighters discuss how to approach the accident.

**Company Officer** (speaking into two-way radio): Engine one on scene of a two vehicle wreck. One vehicle is on its side, looks like people still inside. We'll check for injuries and report.

**Firefighter One**: Wow! Looks pretty bad, cap. How about if I check the blue car? I'll just go check them out.

**Firefighter Two:** Yeah, and I think somebody should check the vehicles for fire danger. Want me to do that, cap?

**Firefighter One**: Maybe we should crib that red one.

**Company Officer** (with authority): Hold on everyone. Calm down. Bob, I want you to set the pump in gear, then stretch a line to the red car. Full bunker gear, S.C.B.A. Dave, you bunker up and go check the occupants of both cars quickly. I want a quick triage report. I'll get some more help coming, then bunker up and check the vehicles for fire hazard and stability.

**Firefighter One:** Got it, cap.

**Firefighter Two:** Sounds good.

---

Evaluate this scenario and describe the positive and/or negative content. If you were the company officer, would you have done anything differently?

_______________________________________________________________

_______________________________________________________________

_______________________________________________________________

**Scenario 2:**

**Company Officer** (speaking into two-way radio): Engine one on scene of a two vehicle wreck. One vehicle is on its side, looks like people still inside. We'll check for injuries and report. Big mess! What do you think we should do? Bob, how about if you set up a line? And Dave, why don't you look over those people?

**Firefighter One:** Wow! Looks pretty bad, cap. How about if I check the blue car? Yeah, I'll just go check them out.

**Firefighter Two:** I think somebody should check the vehicles for fire danger. Want me to do that, cap?

**Firefighter One:** Maybe we should crib the red one.

**Company Officer** (interrupts): But I still think we should probably have a hose line. Maybe I'll ask for more help. What should I do first?

Evaluate this scenario and describe the positive and/or negative content.If you were the company officer, would you have done anything differently?

____________________________________

____________________________________

____________________________________

**Scenario 3:**

**Setting:** Scene takes place at a kitchen table. The company officer and two firefighters are discussing what to do for station clean up.

**Company Officer** (friendly): Well, it's Saturday again guys, station clean up time. Last week we got the bay area, so this week we should hit the inside of the station. What do you guys feel like doing?

**Firefighter One:** I could do the dusting and vacuuming, or maybe the bathrooms.

**Firefighter Two:** I don't mind doing the bathrooms.

**Company Officer:** Okay, that leaves the infamous kitchen. I hope B shift took home their leftovers. Did we get everything? Any areas of the station that need special care? Remember, the chief is coming for lunch next shift.

**Firefighter One:** I think that covers it pretty well.

**Firefighter Two:** Maybe I could wash the windows in the dayroom, too. They're getting pretty slimy.

**Company Officer:** Sounds good to me. Thanks and "be careful out there." (laughs, the other firefighter also laughs)

---

Evaluate this scenario and describe the positive and/or negative content. If you were the company officer, would you have done anything differently?

___________________________________________________________

___________________________________________________________

___________________________________________________________

**Note:** Setting and characters remain the same as in the previous scene.

**Scenario 4:**

**Company Officer** (friendly): Well, it's Saturday again guys, station clean up time. Last week we got the bay area, so this week we should hit the inside of the station. Jack, I want you to clean the bathroom. And Bob, wash the windows, then clean the kitchen. I'll dust and vacuum.

**Firefighter One** (mumbles): The bathroom, again.

**Company Officer** (irritatedly): What's that, Jack? Something wrong?

**Firefighter One** (belligerently): Yeah, I hate the bathrooms! You always assign us things to do and never ask for our input. And, you hardly ever do the bathrooms!

**Company Officer** (controlled, but angry): That's not true. I did the bathrooms last week. And, I don't appreciate your insubordination either. Now, get to it! (looks at firefighter two) You got any problem with washing the windows?

**Firefighter Two:** I guess not.

———————————————

Evaluate this scenario and describe the positive and/or negative content. If you were the company officer, would you have done anything differently?

———————————————————————————————————————

———————————————————————————————————————

———————————————————————————————————————

**Scenario 5:**

**Setting:** Scene takes place in a meeting room. The battalion chief and several company officers are in attendance.

**Battalion Chief:** Next item is the new dress code regulations.

**All Company Officers** (moan).

**Battalion Chief:** Hey, give me a chance. (sincerely) The committee that worked on this project really did a good job. Let's show them the courtesy of listening to what they've developed.

**Company Officer One:** But, Chief, why are we messing with the dress code? The one we have now seems to work well enough. It's going to be tough ramming this new one down the troops' throats.

**Battalion Chief** (reasonably, firmly): Look guys, I don't like this changing of our standards either. But, the chief wants a new dress code and he is going to get one. It's within his rights as chief to tell us how we're going to look. The important thing to remember is that the troops are going to be forming their attitudes based on how you present this thing. You can bad-mouth the whole idea, which will give them permission to complain and moan about it too, or you can present the facts and support management's perogative to implement a different dress code. Remember, you are management too. One of your jobs is to facilitate the implementation of staff policies in a constructive manner.  That's what I'm doing right now, and that's what I expect you to do with your crews. Now, what can I do to make your job easier?

Evaluate this scenario and describe the positive and/or negative content. If you were the company officer, would you have done anything differently?

_____________________________________________________________

_____________________________________________________________

_____________________________________________________________

# REFERENCES — MODULE II


Bennis, Warren & Nanus, Burt. *Leaders: The Strategies for Taking Charge.* Harper & Row, 1985.

McGregor, Douglas. *The Human Side of Enterprise.* McGraw-Hill, 1960.

McGregor, Douglas. *The Professional Manager.* McGraw-Hill, 1967.

# MODULE III

## A SYMBOL IS A PROMISE:

## ORGANIZATIONAL DYNAMICS

"To understand what is happening today
or what will happen in the future, I look back."
— Oliver Wendell Holmes

# PRE-PROGRAM ACTIVITY 1

## ORGANIZATIONAL ATTITUDE QUESTIONNAIRE

1. What is your concept of the mission of a fire protection agency? Define what you see as the goals and objectives of your agency in achieving the mission of fire protection.

2. What do you think is the public's attitude toward the fire service system? How do you think this attitude effects the policy of a fire department?

3. How would you define "professional?" Do you think the fire service is on par with other "professions," such as law, engineering, medicine, and social sciences?

4. What positive or negative forces do you think are at work in society that effect the fire service and its personnel?

5. What would you say are the characteristics of a "good" fire department?

6. What would you say are the characteristics of a "bad" fire department?

7. How do you define "tradition." What does tradition do for you and the organization?

8.  What does the term "teamwork" mean to you? How is teamwork developed?

9.  What does the term "technological obsolescence" mean to you?

10.  What is the most important thing about your job in the fire service?

11.  What's the least important thing to you about your job in the fire service?

12.  What do you think your superior has a right to expect from you?

13.  What do you think your subordinates have a right to expect from you?

14.  What do you think you, individually, have a right to expect from the fire department?

15.  Do you feel you have been given sufficient training opportunities to become competent in the fire service? What new opportunities and materials would you like to see added?

16.  What is the most important program we engage in as it relates to fire protection? Why?




17.  What is the least important program we engage in as it relates to fire protection? Why?

# MODULE III NOTES

# Man's mind, stretched to a new idea, never goes back to its original DIMENSION.

Oliver Wendell Holmes

**This page intentionally left blank.**

# MODULE III

# A SYMBOL IS A PROMISE: ORGANIZATIONAL DYNAMICS

## OBJECTIVES

To provide the viewer and participant with:

- a basic understanding of the process of change

- a basic understanding of the changing requirements for fire officers
- an assessment of their attitudes and orientation with respect to change.

## OUTLINE OF KEY POINTS

**Society and the Fire Service** — The fire service is a function of civilization. Society has brought about the majority of changes in the fire service over the last 25 years. We have and will continue to reflect the values of society.

**Transitions** — Society is always in transition. There is no guarantee that any specific value, technology, philosophy or technique will continue forever.

**Eras of Change** — Some say the fire service is two hundred years of tradition unhampered by progress.

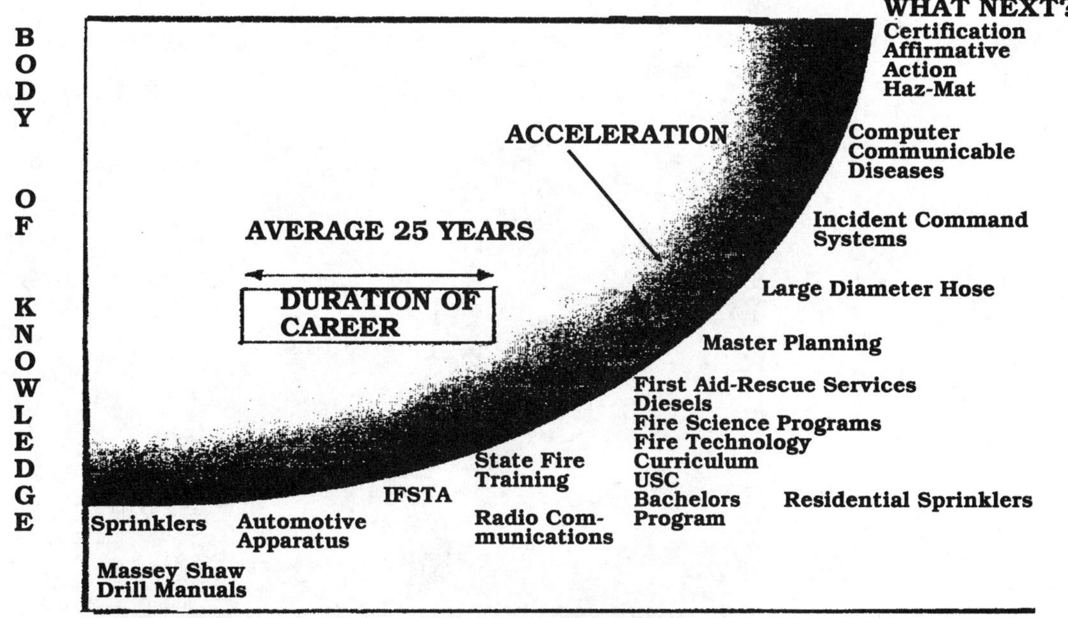


**Figure 3-1   Eras of Change Model**

Time reveals the opposite to be true. We have gone through an extensive evolution of technological change that parallels society in general. Although the fire service has moved along with society in accepting changes, in general it has been a few years behind societies' changes. Currently, most of the changes occurring in the fire service are a result of hazardous materials issues. The model illustrates the fact that we will continue to see change in the future.

**Changes in Job Requirements** — There are two dimensions to changes in job requirements. The first dimension is caused by changes in technology and methodology (steamers vs. diesels). The second dimension is caused by changes in job scope (an engineer vs. a fire officer).

**Levels of Responsibility**— There are various levels in an organization with respect to change. The fire officer is an integral part of the planning process to deal with change.

**Decision Gap** — This model illustrates the common discrepancy between the following factors:

- **Information/Time** — There is a relationship between the sequence of events and the amount of information you have at any given point in time.
- **Defining Your Problems** — This involves accuracy in understanding exactly what you are dealing with. For example, we see this in structure fires when we have heavy smoke. We have a fire in a dwelling place,

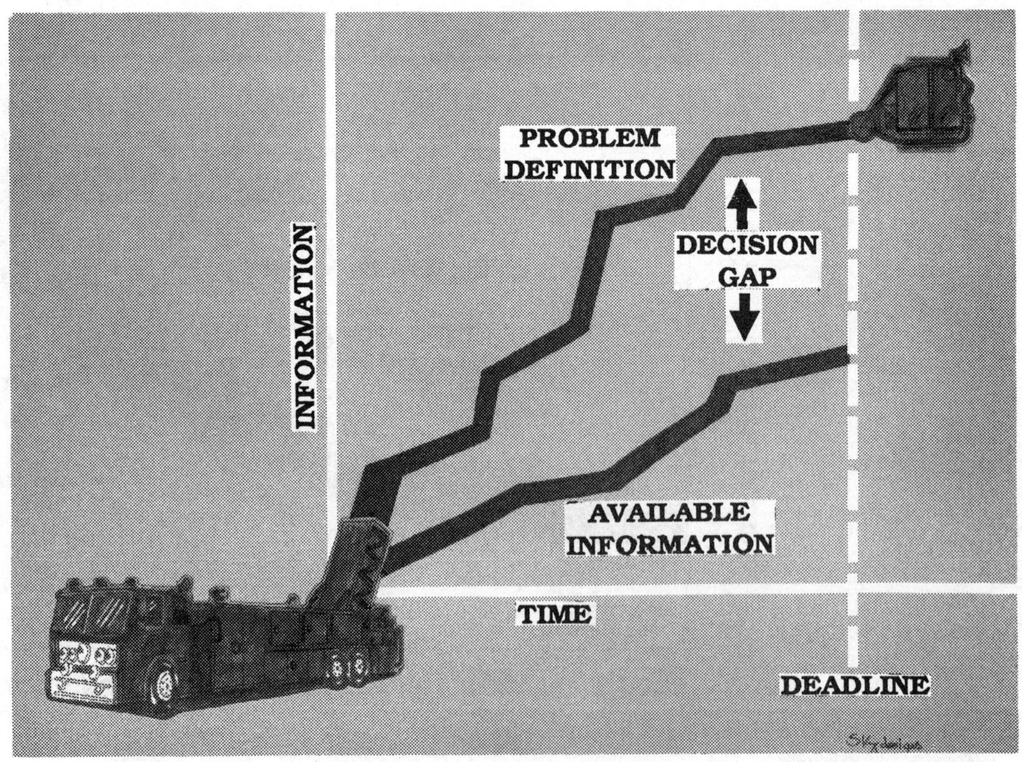


**Figure 3-2   Decision Gap**

but where is it? We have to ventilate to find the fire. In the field of human dynamics, we often have difficulty locating and defining what we are dealing with.

- **Available Information** — This is the total sum of what we actually know about a problem. There is often a gap between what we know to be our problem and what we know about dealing with the problem.

**Skills Needed Model** — The following are skills that need to be developed in order to be an effective leader:

- **Technical** — The ability to complete tasks that are technologically oriented, i.e., how to operate a piece of equipment.
- **Human** — The ability to understand human behavior and communicate at verbal and non-verbal levels.
- **Conceptual** — The ability to understand concepts and relate them to the real world of people and things.

**Process of Change** — There is an actual process for change to occur. It occurs for both individuals and organizations. It is measurable and definable. The fire service has never experienced overnight changes. It has always been a process that involves:

- Acceptance
- Adoption
- Implementation
- Obsolescence.

This process of slow change in the fire service has caused it to be viewed by society as one of the most stable government agencies.

**Input/Output** — Change does not occur in a vacuum. There are inputs, such as events, that lead up to changes and there are outcomes from the process.

**Change Model** — This model (Figure 3-4) illustrates the growth that an individual or organization goes through in adapting to changes. There are two dimensions to the model. They are difficulty and time.

- **Difficulty/Time** — There is a relationship between the degree of difficulty in adapting to change and the amount of time it takes to make the change occur.

**Knowledge Base** — All change begins with a change in information. One must learn about a new way of acting, thinking, or behaving before they can incorporate it into their way of doing things. However, just knowing

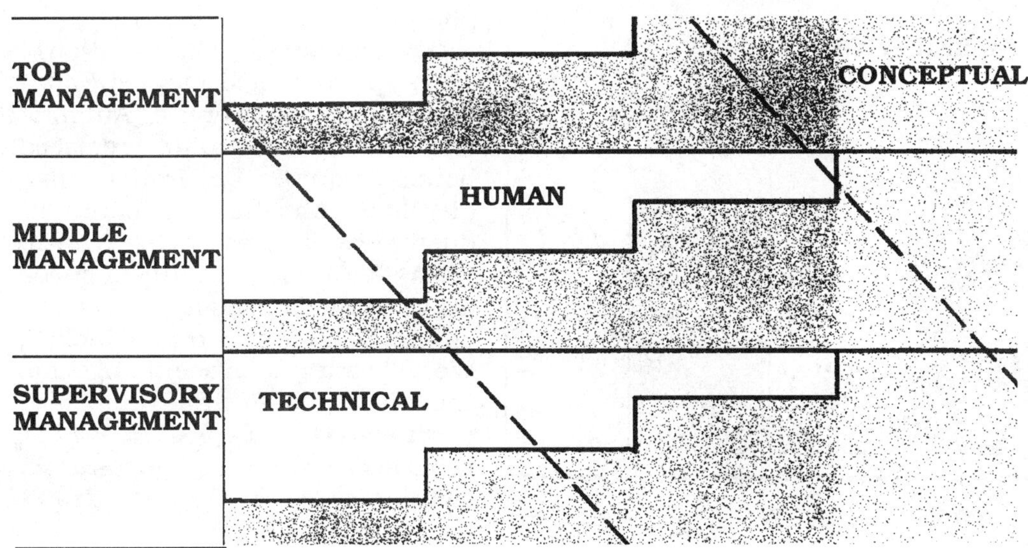


**Figure 3-3   Skills Needed Model**

about something is not what causes the new knowledge to be put to use. It is understanding the concept or idea that allows us to begin to use the new knowledge in our daily lives. Individuals, such as inventors and creators, are constantly forcing our body of knowledge to change with new innovations and concepts.

**Change in Attitude** — One has the option of discarding information regardless of its value. Knowledge is only useful if it is accepted. The first step is to develop an attitude that is open to accepting change. You must be willing to ask, "How can this help?" and "How can I implement this new knowledge?" An adjustment in attitude not only gives individuals permission to learn, but permission to use what they have learned. If we give an individual in the organization access to new information, then we must be willing to let them use what they have learned to change the group behavior within the organization to make it better.

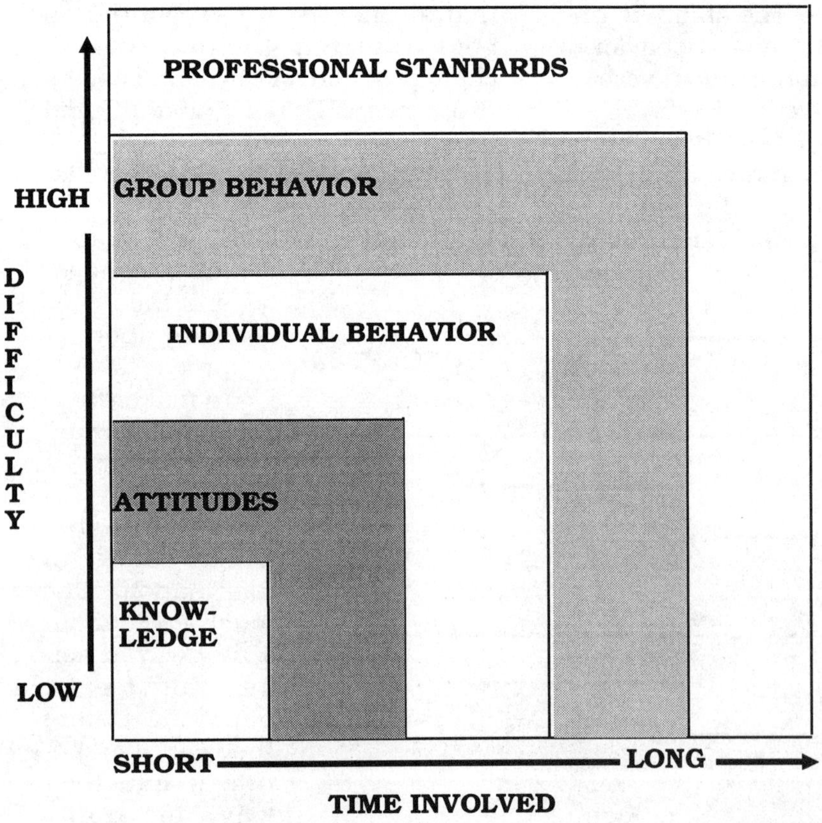

**Individual Behavior** — The outward acts displayed toward the use of new knowledge. This could also be called a "testing period" for a new idea.

**Group Behavior** — The collective act of a group of individuals or an organization in utilizing new information. This might be called the "acceptance period." Changing group behavior takes much longer than the process of gathering new knowledge.

**Professional Behavior** — The standards by which an entire group of specialists utilize new information. This might be called the "ethical period." Changing professional behavior involves changing the standards within the profession. This, understandably, is a very slow process.

**Origin** — The period of initial development for fire service technology, methodology, and philosophy.

**Development** — The evolution of society has created parallel evolutions in the fire service. Automobiles were invented and we replaced horses on our fire apparatus. When diesels were made available to industry, we put them in fire trucks.

**Rapid Change** — Future shock exists in the fire service. The acceleration of change is on the incline. We will continue to see change into the foreseeable future.

**Information Explosion** — The rapid acceleration of changes in society as a result of the electronic revolution.



**Figure 3-4   Change Model**

**Innovation and Adoption Curve** — The distribution of attitudes towards change. Both individuals and organizations can be found on this chart.

**"The value of an idea lies in the using of it."**
**— Thomas Alva Edison**

**State of the Art** — The "cutting edge" of change. That portion of society which is in the forefront of change and progress.

**Innovators** — The knowledgeable with a competitive edge motivated by the desire to achieve economic advantage. Innovations are usually brought about by these people within any profession. They have worked with the system and have an idea of how it would work better.

**Early Adoptors** — The curious with a desire to get ahead and motivated to achieve professional or political advantage. These people are on the cutting edge of change. They do not mind taking risks.

**Obsolescence** — The backwash of society. That portion of society in a resistance role.

**Late Adoptors** — The timid with a desire to never be wrong based on their fear of accountability. New ideas must be proven before they will implement them.

**Resistors** — The uninformed with a desire to maintain the status quo based on either apathy or ignorance.

**Refusers (Extinction)** — The incapable who live in the past and will be replaced as they die off. They literally refuse to accept what is taking place, usually because of a lack of knowledge. These people often become extinct if they do not learn to accept change.

**Mainstream** — The vast majority of society. This portion of society lives with technology and adapts it to daily life. To an extent the fire service must be a mainstream organization. The nature of the fire service requires that the stable base of knowledge used in daily

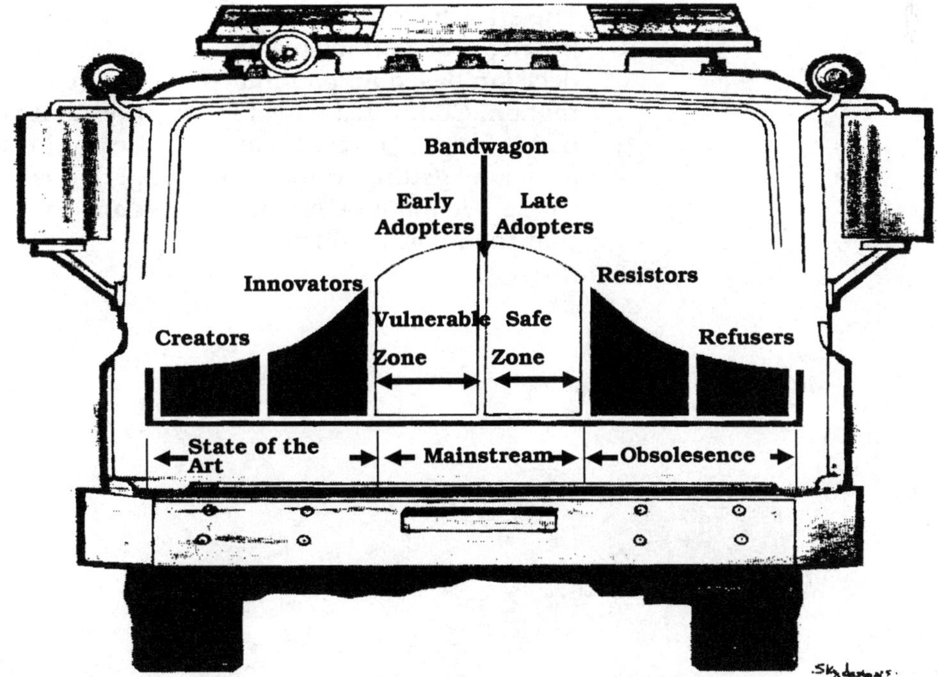


**Figure 3-5  Innovation/Adoption Curve**

operations remain the same. However, we must also keep pace with what is going on in society. We should constantly be questioning what we know and how it can be improved. The mainstream can not remain at a fixed point without becoming stagnant.

**Bandwagon** — The practical who adapt to the real world based on personal and social needs to conform to standards. These individuals prefer to play it safe.

**Risk Zone** — The area of experimentation. This is the "bleeding edge" of change. It is where technology or methodology is the most vulnerable. These individuals look at change as a challenge.

**Safe Zone** — The area of standardization and conformity. This is the foundation for the bulk of activity.

## Coping With Rapid Change

We should each become familiar with the following terms. These terms will assist in understanding and coping with rapid change:

**Technical Obsolescence**— When an object or a process no longer serves the needs of its user group. It is a function of being made useless by a new technology or being abandoned by its user. Two ways to avoid technical obsolescence are:

- **Deviation Amplification** — The error factor that is built in once a person makes a decision based on bad information. Continued use of bad information results in decisions getting worse and worse. We must be certain that the information we are basing our decisions on is accurate and up-to-date.

- **Professional Curiosity** — The quality a person has that requires them to explore for new understanding. A leader must develop this quality if they are to keep their organization on the cutting edge of society.

**Information Half Life** — The time it takes for fifty percent of a knowledge base to be replaced with new knowledge. It usually takes about five years for new information that you hear or read about to be on the scene. This five year time period is called the ***strategic horizon.***

## FUNCTION OF A FIRE OFFICER

As almost everyone knows, an officer has nothing to do except:

- decide what is to be done
- tell somebody to do it
- listen to reasons why it should be done by someone else or why it should be done a different way
- follow-up to see if it has been done
- discover that it has not been done
- inquire why it wasn't done
- listen to excuses
- follow-up again to see if it has been done yet, only to discover that it has been done incorrectly
- show how it should have been done
- wonder if it isn't time to get rid of a person who can't do a thing right
- realize that person has a spouse and family depending on them and that any replacement would probably be just as bad, if not worse
- consider how much simpler the incident would have been if one had done it in the first place
- reflect sadly that the request could have been done correctly in 20 minutes, but you've spent two days finding out why it has taken three weeks for somebody else to do it wrong.

## A SYMBOL IS A PROMISE

Is nothing sacred anymore? No matter what direction we turn today there is a criticism or comment leveled at fire protection. It often seems that everything is in a state of flux. Fire trucks aren't red anymore! Women in the fire service! Fire hose keeps getting larger in diameter and fire trucks are getting more expensive.

Almost every time a group of firefighters get together to discuss this phenomena, someone will raise the issue, "They are destroying all of the traditions of the fire service." Are they really? What are the true traditions of the fire service?

Tradition was once defined as anything that we continue to do after we have lost the original reason for doing it. In that context, one might say that we have a lot of "traditions" in the fire service evidenced in the way we do business. Years of repetition often build up layers of insulation against logic. So it is true that we are a traditional organization.

Perhaps it is time for us to sit back and take a look at what we mean when we talk about tradition. As an example, let me use the engine company. In your fire station you have a diesel gulping, chrome clad, electronically voiced, extremely expensive piece of firefighting equipment. It is about as far removed from the hose cart that was in existence at the turn of the century as the Voyager Rocket sent to Saturn was from Kit Carson.

Yet, when someone decides to build a new fire apparatus that is radically different from the ones that are being advocated by apparatus manufacturers, they are attacked for being "non-traditional." Take attack pumpers, for instance. Those communities that have not opted to use them criticize their lack of effectiveness and challenge their right to be used as part of a firefighting arsenal. Yet attack pumpers put out thousands of fires on a daily basis.

### You Promised To Deliver

*I believe that allowing the erosion of symbolism in the fire service is a mistake.*

We shouldn't lose sight of the fact that there is a difference between a symbol and a tradition. A symbol is a promise to deliver. The American flag is the symbol of freedom. That piece of brightly colored cloth has never won a battle, conquered a territory, repelled an enemy, or passed a law. Yet the American flag is symbolic of our way of life. When we see that flag flying in front of our schools and public buildings, it means that we promise to do our best to provide for the freedom of the individual living under that flag.

So it is with fire service symbols. Take a look at that piece of metal called a badge. Badges don't enforce fire codes. Badges don't risk their lives. They are merely pieces of metal, but they are symbolic. What does your badge stand for?

Many professions use symbols as an expression of their professional demeanor. Next time you go to court notice that judges wear black robes. Pilots wear wings on their chest. Generals still wear stars on their shoulders.

It might be time to ask this question, "What are the true symbols of the professional fire service? What do those symbols promise to deliver?" I believe that allowing the erosion of symbolism in the fire service is a mistake. For example, there are many fire chiefs today who have never worn the uniform of their department since assuming their mantle of responsibility. Someone told them that they are now a "businessman" or an "executive." They feel it is unseemly to wear the uniform, badge, and gold braid accompanying the position they have achieved.

Subsequently, in many communities the fire chief has become an obscure bureaucrat, not much different than the rest of those associated with city hall. This practice often filters down to the other officers in the organization and, in many cases, down to the fire inspectors in the Prevention Bureau. It is no wonder that tailboard firemen question whether they should have uniform standards when the officers in the department set no example. It is the same with automobiles. Many fire chiefs have opted to drive an automobile that is unmarked. While this might provide a certain degree of anonymity, it does not contribute to the visibility of the organization.

## What Kind Of Stuff Are You Made Of

I would like to ask you to participate in a little soul searching. Let me start by asking a very straight forward question. What do you stand for? Do you have a philosophy of fire protection that guides every decision? Could a total stranger be able to determine what you are practicing as a profession? If someone were trying to characterize your performance by placing a symbol on it, what would they choose? Just as football teams are named after objects, would you be characterized as a tiger or a pussy cat? Are you a rock or quicksand? What do you do that epitomizes your profession as a fire officer? Are you proud of your role in society? Do you deliver to your community what you are capable of delivering or do you deliver only what they ask?

Heavy stuff, right? We receive our officer's badge without a set of operating instructions. We spend the better part of our first four or five years as an officer trying to determine what we should be accomplishing.

A friend of mine once suggested that every person who is appointed as a fire officer in the United States should be issued a reproduction of the fire helmet used in 1776, when this country was first formed. The reproduction would serve as a symbol of service and hang in the office as a constant reminder to the new officer. It would symbolize his/her responsibility for providing an ever increasing level of protection to the community against fire and other calamity.

Silly you say? Have you ever noticed that doctors still use an ancient Greek symbol to represent the medical profession? They proudly display that little symbol on their automobile and other places to tell everyone they are a physician. This ancient symbol, however, doesn't keep doctors from embracing new and sophisticated technology designed to keep us living longer.

If we have any traditions in the fire service, they should be traditions that represent a philosophy. It should be traditional that we have courage, dedication, sincerity, integrity, and all the other attributes that make Saturday morning cartoon heroes come alive.

The tools that we use should not be our tradition, but rather it should be traditional that we fight

*It should be traditional that we have courage, dedication, sincerity, integrity, and all the other attributes that make Saturday morning cartoon heroes come alive.*

fire in a certain way. It should be the role of the fire officer to question whether his tools are doing an adequate job, rather than defend the tool simply because it exists. Those of us who are in the profession of firefighting should be forcing changes upon the fire service, rather than waiting for others to insist we change. Looking back at our past, we should recognize the necessity for research and development to improve the tools of our trade.

## The Challenge Of The Future

Currently, we face some of the most explosive changes that the fire service has ever had to cope with. We are entering the computer age. It is going to take some dynamic leadership in order to bring the fire service into this new age with a minimum of impact on its role in society.

There is nothing wrong with us changing the way we do business, as long as we stay in the same business. As a modern fire officer, it is up to you to accept the responsibility for our future. The shape of a fireman's helmet, the color of fire trucks, or the use of the firefighter's axe must be placed in their proper context as being symbolic and not traditional. The trappings of our profession should not limit the acceptance of change.

Just like the trumpets which adorn a company officer's lapel,

*Our trumpets are symbols that should be viewed as promises to deliver to the people of our communities the greatest degree of protection of life and property that is technologically possible.*

Once again, thinking back to our predecessors, we can look to people like James Braidwood of Edinburgh, Scotland and Sir Massey Shaw of the London Fire Brigade, who literally forced the fire service into some of the first training programs. How about James Gulick of New York City, Keith Klinger of Los Angeles County? In our past we have had passionate and involved leadership that has brought the fire service forward by leaps and bounds in the protection of life and property.

these "traditions" stand for something far greater. Our trumpets are symbols that should be viewed as promises to deliver to the people of our communities the greatest degree of protection of life and property that is technologically possible. In that context, our symbols should be treated with respect and worn with pride!

The Greek god, Prometheus, gave man fire and was condemned to an eternal punishment because he had given mankind a tool that would forever change its relationship with the world. If the ancient Greek gods saw fit to punish Prometheus for giving man fire, imagine how much respect they would have for those of us who have been given the responsibility of protecting society against fires misuse and its destructive capability.

86

## NEW VOCABULARY ENTERS THE FIRE SERVICE

When was the last time you used a fountain pen? You remember fountain pens, don't you? They were those devices that were cleverly designed to fail at the most inopportune time. Inside the pen was a small rubber bladder that was filled with ink from an inkwell. When you were forced to use them, one of two things usually occurred. They either splattered ink all over the paper as the mechanism malfunctioned or leaked ink into the breast pocket of your new white shirt. In spite of these inconveniences, fountain pens were used extensively for several decades to record the written word. They were better than pen quills, but not good enough because their flaws led someone to invent a better mechanism.

The new mechanism was the ballpoint pen. Although the ballpoint is still popular, it has spawned a new generation of writing mechanisms called felt tip pens. The evolution of the mechanism we use to put words down on paper continues to change.

### What's Happening To Our Technology?

Now what has that got to do with the fire service? It illustrates a concept that is in continual operation in the world around us and impacts the fire service. The concept is called technological obsolescence. Simply stated, ***technological obsolescence*** occurs when something no longer has its original utility because technology has brought about a replacement that is more effective, less expensive, or has some virtue that makes it more desirable than the original product. Technological obsolescence turned the quill pen into a museum piece.

Technological obsolescence is operating in the field of fire protection. Can you think of an example of something that was important to you when you first became a firefighter that has been replaced since your career began? I can recall hearing lectures on the use of canister gas masks in my recruit training. Today canister gas masks are not only obsolete, they are outlawed for the kind of environment in which a firefighter is required to function. There are many other examples.

Someone once said that if a fire chief who died in the 1880's was resurrected and dropped in the middle of a modern fire station, it wouldn't take him long to figure out which end of the fire truck the water came out of. We have embraced concepts in the fire service for over one hundred years that have their fundamental roots in the design of fire equipment at the turn of the century. For example, the 2 1/2" discharge outlet, the 1,000 to 1,250 GPM pump, or the straight board nozzle have remained unchanged over this past century.

What if that resurrected fire chief was dropped behind the desk of a fire inspector doing record keeping on a computer? Would he recognize the computer as being similar to the old station log book? What if that ghostly fire chief was required to tag along with his modern counterpart to attend a subdivision meeting and discuss the intricacies of environmental impact reports and fiscal impact analysis. It is obvious that technological obsolescence is occurring in our business also. It's a term that is now as much a part of the firefighter's vocabulary as "size up."

One term spawns another. For example, technological obsolescence implies that as something becomes obsolete, it is replaced by something that is not only newer, but more complicated, more sophisticated, or different to operate. That leads to the second term that we would like to discuss, information half life.

## What's Happening To Our Information?

*Information half life* is the period of time it takes for a certain amount of information to become obsolete. For example, can you imagine how different it is today to be an engineer designing calculators than it was to be an engineer working at Burrough's Company in the 1920's. If an individual graduating from college with an electronics major in 1940, did nothing to keep himself abreast of his field, his knowledge would have been outdated by 1945. This is the length of time it took for at least 50% of the electronics knowledge to change and be replaced. This is information half life in operation. It is the period of time it takes for 50% of what you know to become inadequate or inaccurate.

At one time, information half life in the fire service was extremely long. There was little change in the methods of firefighting from the late 1700's until the late 1800's. Compare that with the rapid changes that are going on in the field of equal opportunity employment. Consider the rapid changes in the funding structure of fire departments. In addition, recall the rapid increase in the use of electronic technology in the fire service. In other words, if an individual receives his education in fire protection technology, that expertise cannot remain frozen. The horizons continue to change almost daily in the fire service. As technology becomes obsolete, so does our information. As our information

becomes obsolete, so does our education.

How do these two new terms relate to you as a fire officer? First and foremost is the idea that if equipment can become obsolete, so can people. One of the roles of a modern fire officer is to be engaged in active curiosity to assure that he is using the most up-to-date methods and equipment to perform his job. If a person can become obsolete, so can his contribution to the organization when his knowledge is not sufficient or current.

That leads to the second implication, we can never stop learning! No one ever knows all of the answers. As a matter of fact, it is almost a rule of thumb that the more you explore an occupation, the more you find that your knowledge is limited. You will not be able to rely on your own personal experience and education to acquire the knowledge you need to do an adequate job. You will need to seek out the knowledge of others who have experience in areas that you do not.

There is a term for this process which is new to the fire service. It is called executive curiosity. *Executive curiosity* primarily deals with a concept of never being satisfied that what you know is all you need to know. It is that nagging feeling that continues to cause a person to ask why! Contrary to the two previous terms we discussed, executive curiosity is not an event, it is a process that is ongoing almost continuously.

*As technology becomes obsolete, so does our information. As our information becomes obsolete, so does our education.*

Generally speaking, individuals who have a high degree of executive curiosity have a very low exposure to information half life and technological obsolescence. Continued inquisitiveness into the changes that are going on in the world continues to feed new data to the individual who has the responsibility for keeping his fire department afloat in a sea of change.

Executive curiosity has an additional advantage. It helps reduce the seriousness of the mistakes that we make, and we all make mistakes from time to time! There is no sin in an error. The only sin is failing to realize that the mistake is taking us in a direction where we can suffer serious consequences.

That leads to another new word for the fire officer's vocabulary. It is called deviation amplification. One of the best examples of **deviation amplification** can be found in looking at the control of a rocket. When a rocket is fired from its launching pad and begins to go into the upper atmosphere, there are different guidance systems that are used to control the thrust and the trajectory of the rocket. If the instruments are a bit off, they can cause the rocket to be off at least five degrees. This small error can cause the rocket to miss its target by hundreds of miles. As little as a one degree error can cause the rocket to totally miss its intended target.

If the instrumentation in the rocket is malfunctioning and attempts are made to correct the five degree error, it can inadvertently go five degrees to the other side. The consequences of errors in rocket control become quite serious. Imagine that the computer control instruments lock into a new, but incorrect, ten degree deviation. Attempts to correct the rocket by going ten degrees back to the right will cause the rocket to be off course now by twenty degrees. In some instances, the individuals responsible for mission control may realize that deviation amplification is beginning to occur and push the button calling for the rocket to self-destruct.

## Is Your Ship On Course?

**Deviation amplification** occurs when an error is allowed to continue. The longer the error goes on, the farther we go from our intended course of action. Now, let's imagine that instead of a rocket escalating out of control in the upper atmosphere, we have a methodology or an idea that is leading us in the wrong direction. Do we have the courage to push the destruct button to make sure that the deviation is not amplified to the point where it destroys something beneficial? As a fire officer it is important that we maintain surveillance of what we are attempting to accomplish. We need to be certain that if deviations occur we do not allow the error to repeat endlessly, thereby amplifying the consequences in the future.

As I mentioned earlier, these new words do not sound like firefighting jargon. However, it is necessary that as our world changes, we need to update our vocabulary. For years the jargon of the of the fire service would serve us in almost any conversation with our peers from city to city. We could talk about deuce and a half or triple combinations. Now the board room has changed. The four words that were added to your vocabulary — technological obsolescence, information half life, executive curiosity, and deviation

> *Individuals who have a high degree of executive curiosity have a very low exposure to information half-life and technological obsolescence.*

amplification — are part of a new vocabulary that will be important to the future of fire protection. Don't make the mistake of assuming that they do not apply to you merely because you have never heard or used them in your conversation. They are merely indications of a changing world in which the professional fire service is learning to cope.

Stop and think about these words for a second and see if you can find specific examples in your own personal experience that reflect these concepts. Our way of doing business has changed considerably in the last twenty years. New programs have come on the horizon. Other programs have slipped into anonymity. Some programs have been modified to the point where their original designers would have difficulty recognizing them.

We can be certain that this phenomena will continue to occur for as long as there are people involved in the decision making process. We are not the same today as we were yesterday and we will not be the same tomorrow as we are today. Recognizing the significance of this phenomena allows us to be a part of the changing process rather than simply reacting to the changes. It does not prevent changes from occurring, rather it allows the fire service to participate in the shaping of it's destiny.

**"There are two kinds of fools. One says, 'This is old, therefore it is good.' The other says, 'This is new, therefore it is better'."**

**- Dr. Laurence J. Peter**

# POST-PROGRAM ACTIVITY 1

*Instructions:* Indicate with an "X" whether you agree or disagree with the following comments.

|  | AGREE | DISAGREE |
|---|---|---|
| 1. This is just a job. | _____ | _____ |
| 2. Most of our problems are caused by lack of funding. | _____ | _____ |
| 3. Most of our problems are caused by personalities. | _____ | _____ |
| 4. We often change just for change sake. | _____ | _____ |
| 5. I'm better at this job today than I was a year ago. | _____ | _____ |
| 6. My ideas are often suppressed. | _____ | _____ |
| 7. I perform as well as I can most of the time. | _____ | _____ |
| 8. There is no room for improvement in the fire service. | _____ | _____ |
| 9. I know my job and I think I do it as well as anyone else. | _____ | _____ |
| 10. I'm satisfied with my role and responsibility in the department. | _____ | _____ |
| 11. I'm not satisfied with my role and responsibility. | _____ | _____ |

|  | **AGREE** | **DISAGREE** |
|---|---|---|

12. I'm concerned about the future of the fire service. _____ _____

13. I'm concerned about the future fire problems for our city. _____ _____

14. Things are moving too fast for me. _____ _____

15. We ought to go back to the way things were done when I was hired. _____ _____

16. Major policy changes in our department should be subject to a "majority vote." _____ _____

17. I feel insulated from the decisions made in the department. _____ _____

18. Our department is not as dynamic as it was a few years ago. _____ _____

19. Most of the people in our department want to be "hand fed" information. _____ _____

20. Most of the personnel in our department want to be involved more than they are. _____ _____

21. At this time everybody in our department is fairly equal in knowledge and experience. _____ _____

22. At this time, there is a considerable difference in the range of knowledge and experience in the department. _____ _____

|  | AGREE | DISAGREE |
|---|---|---|
| 23. Compared to my previous job (before you became a firefighter), this job is more satisfying. | ____ | ____ |
| 24. Compared to my circle of friends (non-firefighters), my job is considered more important than their occupations. | ____ | ____ |
| 25. If I had to choose over again, I would choose to become a member of the fire service. | ____ | ____ |

# POST-PROGRAM ACTIVITY 2

## ATTITUDE QUESTIONNAIRE

*Instructions:* In terms of attitude and feeling about yourself and the fire department, how do you react to the following statements?

|  |  | YES | NO |
|---|---|---|---|
| a. | Do you want stronger supervision by your superior officers? | ____ | ____ |
| b. | Do you feel you understand your role in the department? | ____ | ____ |
| c. | Do you feel you understand the direction of the department? | ____ | ____ |
| d. | Do you feel you have been evaluated fairly? | ____ | ____ |
| e. | Do you ever feel frustrated? | ____ | ____ |
| f. | Are you ever afraid to ask "why"? | ____ | ____ |
| g. | Do you feel you have a good future in fire protection? | ____ | ____ |
| h. | Do you feel there are bottlenecks in the department? | ____ | ____ |
| i. | Do you feel you are motivated? | ____ | ____ |

# POST-PROGRAM ACTIVITY 3

## CHARACTERISTICS FOR SUCCESS

Indicate how important the following characteristics are to you.

|  | | **LEVEL OF IMPORTANCE TO YOU:** | |
|---|---|---|---|
|  | | **VERY IMPORTANT** | **NOT IMPORTANT** |
| a. | Having personal security | ____  ____ | ____  ____ |
| b. | Facing challenges | ____  ____ | ____  ____ |
| c. | Having confidence in myself | ____  ____ | ____  ____ |
| d. | Having credibility with others | ____  ____ | ____  ____ |
| e. | Having empathy | ____  ____ | ____  ____ |
| f. | Opportunity to prove myself | ____  ____ | ____  ____ |
| g. | Being productive | ____  ____ | ____  ____ |
| h. | Sense of humor | ____  ____ | ____  ____ |
| i. | Taking pride in my work | ____  ____ | ____  ____ |
| j. | Being creative | ____  ____ | ____  ____ |
| k. | Being proactive | ____  ____ | ____  ____ |
| l. | Getting consensus | ____  ____ | ____  ____ |
| m. | Having freedom | ____  ____ | ____  ____ |
| n. | Being in "balance" | ____  ____ | ____  ____ |
| o. | Personal growth | ____  ____ | ____  ____ |
| p. | Physical fitness | ____  ____ | ____  ____ |

**LEVEL OF IMPORTANCE TO YOU:**
**VERY IMPORTANT          NOT IMPORTANT**

q.  Having curiosity          ____  ____  ____  ____

r.  Having integrity          ____  ____  ____  ____

s.  Having openness          ____  ____  ____  ____

t.  Considered sincere          ____  ____  ____  ____

u.  Considered honest          ____  ____  ____  ____

v.  Being happy          ____  ____  ____  ____

w.  Have compatibility          ____  ____  ____  ____

x.  Have understanding of our goals          ____  ____  ____  ____

y.  Having sense of self worth          ____  ____  ____  ____

z.  Pursuing excellence          ____  ____  ____  ____

aa.  Have recognition of deeds          ____  ____  ____  ____

bb.  Knowing the future          ____  ____  ____  ____

cc.  Being involved          ____  ____  ____  ____

# REFERENCES — MODULE III


Cetron, Marvin & O'Toole, Thomas. *Encounters with the Future: A Forecast of Life into the 21st Century.* McGraw Hill, 1983.

Theobald, Robert. *The Rapids of Change.* Knowledge Systems, 1987.

Toffler, Alvin. *Future Shock.* Bantam Books, 1974.

Toffler, Alvin. *The Third Wave.* Bantam Books, 1981.

**This page intentionally left blank.**

# MODULE IV

## PROBLEM SOLVING AND CONFLICT

## RESOLUTION:

## THE BOTTOM LINE

"The only things that evolve by themselves in an organization
are disorder, friction, and malperformance."
— Peter Drucker

# PRE-PROGRAM ACTIVITY 1

## CONFLICT QUESTIONNAIRE

*Instructions:* Proverbs state traditional wisdom. The following proverbs and statements reflect traditional wisdom for resolving conflicts. Read each one carefully. Using the scale given below, indicate how often each proverb or statement describes how you would typically act in a conflict.

5 = Very typical of the way I act in a conflict.

4 = Frequently typical of the way I act in a conflict.

3 = Sometimes typical of the way I act in a conflict.

2 = Seldom typical of the way I act in a conflict.

1 = Never typical of the way I act in a conflict.

___

____ 1. Soft words win hard hearts.

____ 2. Come now and let us reason together.

____ 3. The arguments of the strongest always have the most weight.

____ 4. You scratch my back, I'll scratch yours.

____ 5. The best way of handling conflicts is to avoid them.

____ 6. When one hits you with a stone, hit him with a piece of cotton.

____ 7. A question must be decided by knowledge and not by numbers if it is to have a right decision.

____ 8. If you cannot make a person think as you do, make him do as you think.

____ 9. Better half a loaf than no bread at all.

____ 10. If someone is ready to quarrel with you, he isn't worth knowing.

____ 11. Smooth words make smooth ways.

____ 12. By digging and digging the truth is discovered.

____ 13. He who fights and runs away lives to run another day.

____ 14. A fair exchange brings no quarrel.

____ 15. There is nothing so important that you have to fight to get it.

____ 16. Kill your enemies with kindness.

____ 17. Seek till you find, and you'll not lose your labor.

____ 18. Might overcomes right.

____ 19. Tit for tat is fair play.

____ 20. Avoid quarrelsome people — they will only make your life miserable.

# MODULE IV NOTES

**This page intentionally left blank.**

# MODULE IV

# PROBLEM SOLVING AND CONFLICT RESOLUTION: THE BOTTOM LINE

## OUTLINE OF OBJECTIVES

To provide the viewer and participant with:

- a basic understanding of problem solving models
- a basic understanding of the process and styles of conflict
- a basic understanding of the causes of stress in an organization
- an assessment of how attitudes and orientation effect conflict
- a basic understanding of the skills involved in handling disputes and complaints.

## OUTLINE OF KEY POINTS

**Stress Model** — This model illustrates the relationship between what a person wants and what external influences wish to occur. Outside forces may produce stress within an individual. How much control the person feels they have in altering the stressful forces will determine what the effect will be in

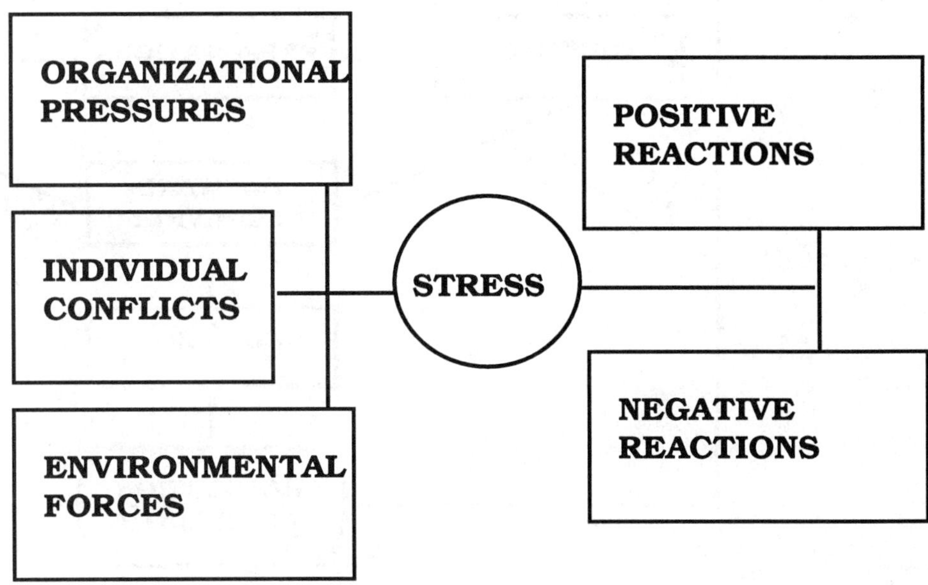


**Figure 4-1   Stress Model**

the individual. Stress can result in one of two types of reaction:

- **Eustress** — A positive reaction that results from the individual understanding that they may be successsul in changing the stressful circumstance. It is also called the "competitive edge." This positive stress is necessary to keep us alert and maintain a heightened awareness of what is going on around us. Effective leadership creates eustress.

- **Distress** — A negative reaction that results in dysfunctional behavior. The individual feels they cannot be successful in changing the stressful circumstances. It is also called "burnout." If distress is caused by someone within the firehouse, it will not go away by ignoring it. Time does not change the situation, instead the stressful circumstances only worsen. Ineffective leadership creates distress.

**Taking Charge of Stress** — No one can create stress within you. You create stress on yourself by your reaction to the situation. The higher your level of personal confidence, the less others will be able to keep you in a stress mode.

**Congruence Concepts** — This model illustrates the relationship between the past, present, and future in achieving cooperation and compliance in an organization. It shows the difference between the expectations of the organization and the expectations of the individ-

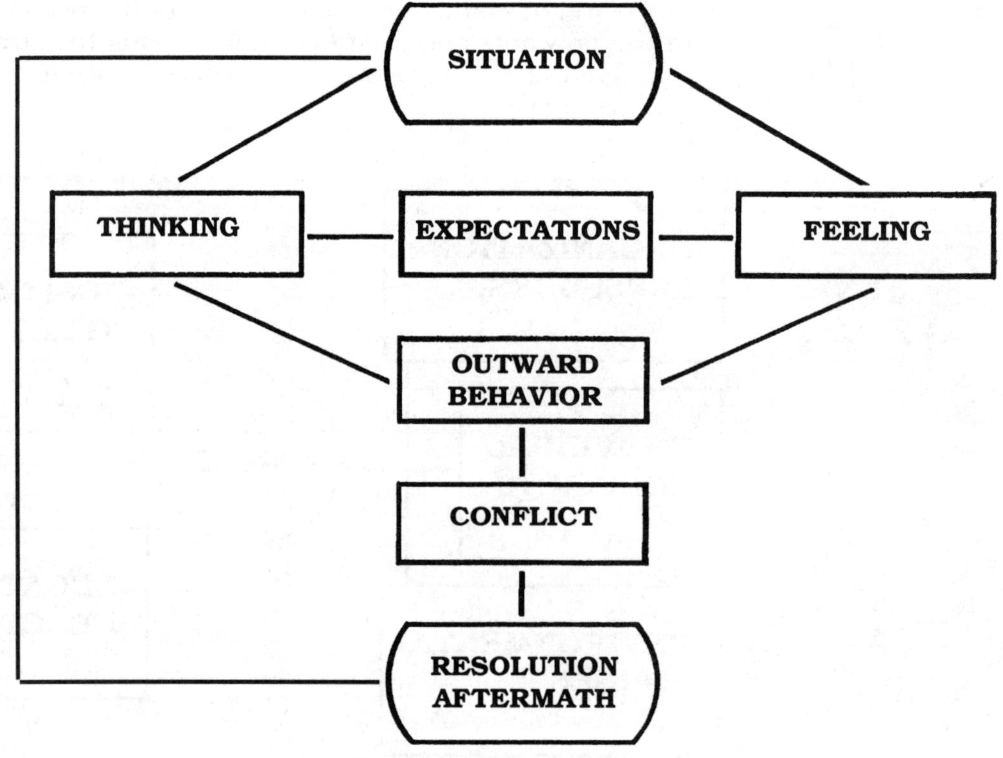


**Figure 4-2   Congruence Model**

ual. It further illustrates that if the process of resolving the conflict has been thorough, then the outcome of resolving the conflict will be achieved more quickly. The process of resolving conflict is as important as the resolution itself. The amount of time you take in resolving conflict will predict whether you are good at it or not.

**Destructive and Constructive Conflict** — Conflict always exists. When it is constructive it builds a better team. When it is destructive it results in confrontation and unnecessary competition. There are different styles of handling conflict. However, the most important concept in resolving conflict is to use both thinking and feeling skills. Most problems tend to be resolved on the feeling side, without the balance of thinking through the conflict or problem. If your outward behavior in helping resolve the con-

flict does not reflect that inwardly you have carefully thought through the issue, then the conflict resolution process will be dramatically slowed down. There are also different styles of conflict (see pages 109-112). A conflict style can be chosen to fit a situation when you understand them. If only one style of conflict is used, you become very predictable. You will find that the Knowledge/Experience Ratio (Figure 1-3) comes into play here. As your conflict resolution skills increase, you will find that you will be able to quickly:

- evaluate the problem
- break the problem down into its component parts
- understand each component by finding out facts
- set objectives
- prioritize which part needs to be dealt with first
- begin attacking the problem.

**Figure 4-3   Win-Lose Model**

**Shift Wars** — In the fire station we have all seen conflict effect daily operations. There is a concept called ***"win-win" or "win-lose"*** (Figure 4-3). The fire officer needs to build cooperation by understanding how conflict can be converted to cooperation. When this is accomplished, all people involved in the conflict can feel they have "won" or learned something from the conflict resolution process.

**Fire Suppression vs. Prevention** — Historically, there has been a difference between line and staff functions. Cooperation in an organization is driven by the identification of common interests. What common interests are shared by prevention and suppression personnel? Where does the fire officer fit in developing understanding in an organization?

**Interpersonal Conflicts** — Conflicts can be created over the smallest items. Developing a mechanism for reducing conflict involves a considerable amount of time in communication. The ***competition model*** illustrates the relationship between communications and the level of agreement in an organization. The leader within the organization can cause the organization to move along the curve, all the way from a lose-lose situation to a win-win situation, depending on the level of control, cooperation, and participation that he/she allows in resolving conflict. Conflict tends to be lowered when the leader allows others to participate in setting goals and decision making.

**Figure 4-4  Competition Model**

**Useful Conflict** — There are definite reasons for people to develop disagreements in an organization. The danger is when a person becomes disagreeable. The corollary to consensus is dissension.

**Stabilization Within the Fire Station** — Developing cooperation in a fire station environment requires an understanding of the realities of fire station life. The fire station environment is effected by long periods of togetherness that can result in either cooperation or conflict. The combination of life-styles and the need to respond to a wide range of emergencies creates stress. The fire officer needs to de-velop skills in basic problem solving and decision making to create a stable atmosphere.

**Basic Problem Solving** — This model illustrates the relationship between knowledge/experience and the need to reinforce your decision when the answer is straight forward and uncomplicated.

**Complex Problem Solving** — This model illustrates the need to evaluate complex problems in terms of objectives and priorities. It introduces the idea that there are alternative ways of achieving satisfactory results in complex situations. This is the model you should use most often in problem

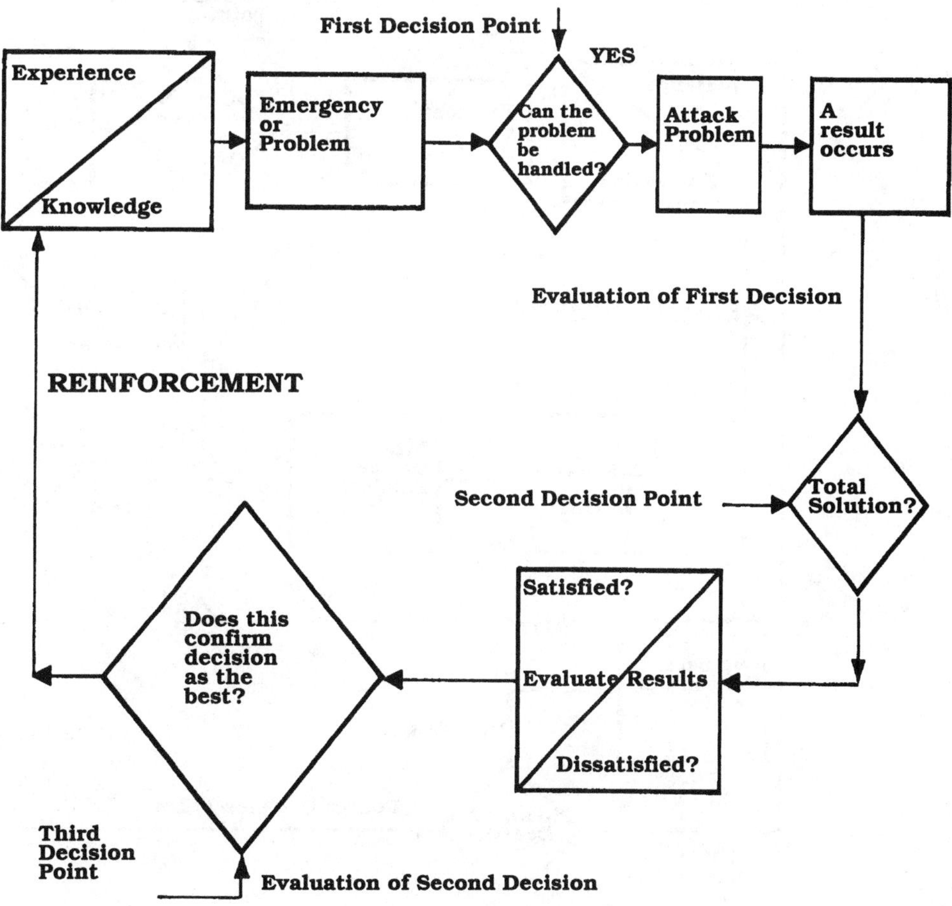


**Figure 4-5   Basic Problem Solving Model**

solving since it causes you to look deeply into what is going on. To use this model effectively, you must continually be increasing your Knowledge/Experience Ratio (Figure 1-3).

**Peer Relationships** — The use of feedback and the concept of sharing expectations is an integral part of reducing stress. Openness and directness as part of a communications process are essential in maintaining working relationships.

**Superior/Subordinate Relationships** — The use of goals, objectives, and priorities is an integral part of eliminating stress between superiors and subordinates. Developing successful conflict resolution within your organization involves two things:

- be careful not to allow individuals to participate in areas where they should not, and
- allow individuals to participate in appropriate areas, especially areas that are meaningful to them. Individuals should be encouraged to be as creative in the firehouse as they are outside in their private time.

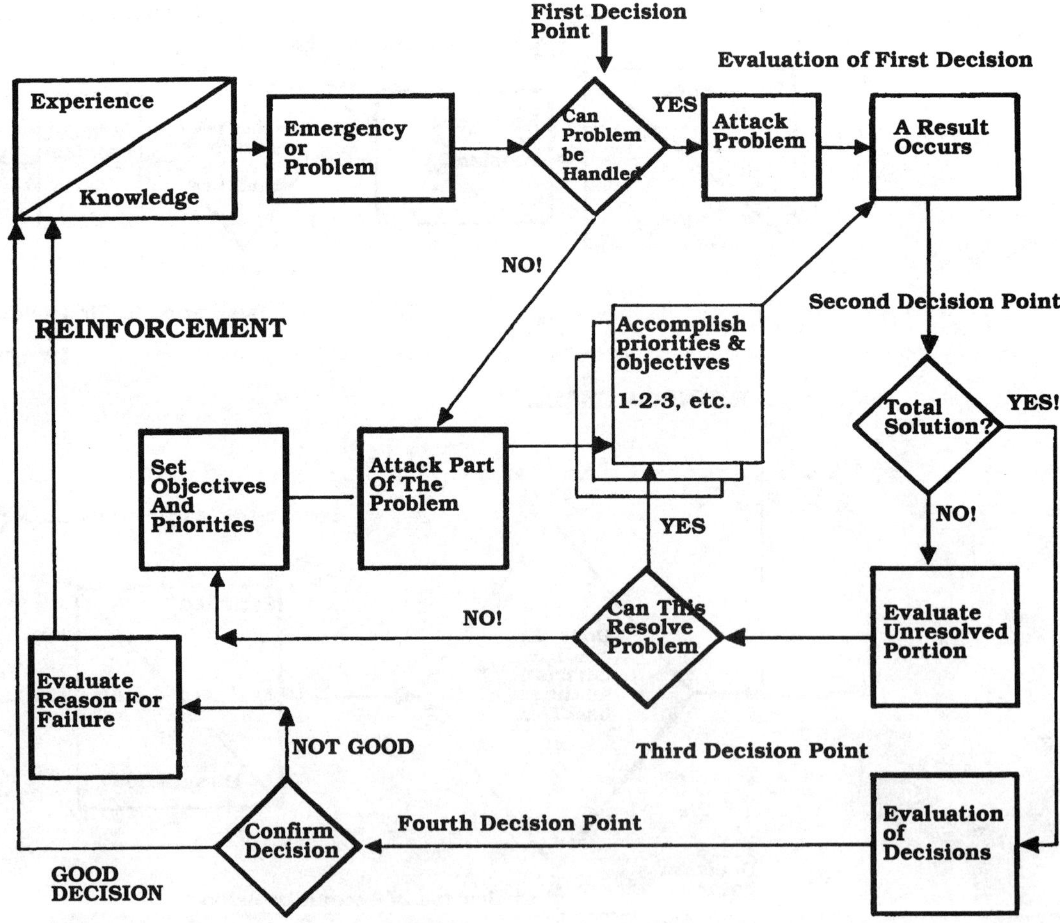


**Figure 4-6  Complex Problem Solving Model**

# STYLES OF CONFLICT

**COMPETING**

| **Situations Competing Style Is <u>Most</u> Useful In** | **Situations Competing Style Is <u>Least</u> Useful In:** |
|---|---|
| 1. Quick, decisive action, e.g., emergencies | 1. Around "yes" people. |
| 2. Important issues, unpopular courses of action, e.g., cost cutting or discipline. | 2. With people who are timid or afraid to admit ignorance or uncertainty. |
| 3. Company policies when you know you are right. | 3. On unimportant issues when consequences may be negative. |
| 4. Protect yourself against others who take advantage of noncompetitive behavior. | 4. When you are unsure of position. |

# AVOIDING

| **Situations Avoiding Style Is <u>Most</u> Useful In:** | **Situations Avoiding Style Is <u>Least</u> Useful In:** |
|---|---|
| 1. When issue is trivial, of only passing importance. | 1. Issues that interfere with work. |
| 2. When you see no chance of satisying your concerns, e.g., company policy. | 2. Wasting time skirting problems that should be addressed. |
| 3. Damage of confronting conflict outweighs benefits. | 3. Issue getting in way of good relationship. |
| 4. To let others cool down, reduce tensions, regain composure. | 4. Resolving problem likely to have positive consequence. |
| 5. Need to gather more information. | |
| 6. Others can resolve more effectively. | |
| 7. When issue really a symptom of something else. | |

# ACCOMMODATING

| **Situations Accommodating Style is <u>Most</u> Useful In** | **Situations Acccommodating Style is <u>Least</u> Useful In** |
|---|---|
| 1. When you realize you are wrong. | 1. When you have good ideas that should be heard. |
| 2. Issue more important to someone other than yourself. | 2. When getting recognition with the boss is important. |
| 3. Build up credits for later more important issues. | 3. When discipline is important and needed. |
| 4. Continued fighting would damage your position. | 4. When policies, procedures, assignments necessary or right for organization. |
| 5. Keeping harmony is important. | |
| 6. Aid in development of subordinates by allowing experimentation. | |

## COLLABORATING

**Situations Collaborating
Style Is <u>Most</u> Useful In**

1. When both concerns too important
to be compromised.

2. Merge insights from different people
to solve a problem.

3. When you want to learn other's
position, test your own.

4. Assume work through hard feelings.

**Situations Collaborating
Style Is <u>Least</u> Useful In**

1. When don't have time to talk
through.

2. When issue is not worth so
much effort.

3. When you have respon-
sibility and don't need
committment.

4. When trust is very low.

## COMPROMISE

**Situations Compromise
Style Is <u>Most</u> Useful In**

1. When goals moderately important.

2. When two opponents have equal
power and committed to exclusive
goals, e.g., labor/management
bargaining.

3. Achieve temporary settlements.

4. Arrive at expedient solutions
under time pressure.

5. Back up mode for Fight or
Win/Win styles.

**Situations Compromise
Style Is <u>Least</u> Useful In**

1. When compromising means
losing too much.

2. When cycle of "games-
ship" undermines work
relationships.

3. When issues or principles
too important.

## CONFLICT RESOLUTION: THEORY AND PRACTICE

### Introduction

Conflict. It's a word we are exposed to in the daily news, at work, even at home. Most of us are constantly dealing with conflict and aren't even aware of it. Dealing with the salesperson who "can't remember" your order or the person cutting into the line you've been standing in may simply seem like the price of daily living. Yet in the truest sense, these are examples of conflict. How we deal with individuals in conflict and how well we feel at the end of our interaction can be influenced by our understanding of the concepts of conflict and conflict resolution. In the following pages you will be exposed to information about the process of conflict — what it is, how it develops, and how you can more effectively deal with it.

### The Individual's Reality and Its Contribution to Conflict

To fully understand conflict, it is important to understand the system within which conflict develops. For us as human beings, the system is usually called "the world." It is the stage for what we think of as "reality." The major concept you need to consider is that reality is defined from our individual perception of the world. Each person has a unique view of what they call reality. They will live and act in the world based on this unique view. As we grow, we learn to use labels (words) to help us maintain a solid, structured view of "what's out there." If you doubt the importance of these labels, think of how differently you feel about "G. Brown, Therapist" and "G. Brown, The rapist." In effect, one space has changed your emotional perception of G. Brown. As mentioned, humans use labels to maintain a sense of structure in his/her reality. Without these structures we lack a focal point. We move into chaos, a state of randomness that most of us find extremely disorienting. Perhaps a "thought experiment" can be used to give you a sense of what a chaotic state would be like.

Imagine you have gone to the store. As you begin to move down one of the aisles, a young woman approaches you and begins to ask for directions to another part of the store as if you were a store employee. As you attempt to correct her, she looks at you, smiles, and continues right on as if you really were an employee. When you again try to correct her, the young woman appears to get irritated and stops a passing employee to complain about your behavior. Thinking that you can now get the misunderstanding cleared up, you explain to the employee what has just occurred. As you finish, the employee looks at you and asks if you're all right. He then suggests that if you are O.K., you might want to help the lady with her request. At this point, stop and imagine what you might be feeling. Confusion? Anger? Fear? All are realistic responses to being thrown into a chaotic situation. You might note that you probably aren't feeling happy, secure, or peaceful.

The supermarket discussion illustrated a conflict of two realities — yours and theirs. Your reality view was contradicted by the reality

> *The major concept you need to consider is that reality is defined from our individual perception of the world. Each person has a unique view of what they call reality.*

view of another. Your reaction was a way of defending your view of "how the world is" and maintaining your sense of a stable, predictable world. Eeverybody sees the world uniquely. Not seeing what we expect to see creates a feeling of insecurity.

To truly understand the function of conflict it is important that we understand what is conflicting. The word "conflict" is derived from a Latin word meaning "to strike against." This implies the striking together of different realities. When we are dealing with irate individuals we tend to label their anger or the dispute as "the conflict." The anger of the moment is the emotional reaction brought on by the conflict, or striking together, of two different realities. As we discussed, when a situation of conflict occurs, each individual attempts to restructure or control the situation to restore their reality.

Should the individual's attempt fail, behavior can escalate from quiet, internal disagreement to verbal disagreement, or even physical disagreement. This escalation usually proceeds in a predictable fashion. In fact, this progression is so predictable that individuals who do not follow it are often thought to be mentally ill. An example would be a person who just "blows up" for no reason without giving any verbal clues.

To summarize:
- Each individual views the world/reality in their own unique way.
- Any situation that upsets this reality view will be reacted to emotionally and/or physically.
- These reactions tend to follow a predictable pattern moving from internal reactions, through verbal reactions, and ending with physical reactions as a means of reestablishing a sense of control or well being.

## Applying Conflict Theory

Although the concepts discussed previously are interesting, they may not seem very practical. Now let's see how we can put the concept of conflict theory to use.

Perhaps the most basic use of the theory is to remind you that the behavior you are observing in an individual in crisis is not **caused** by you personally. When you see anger, rage, frustration, sadness, hopelessness, fear, guilt, or anxiety as a reaction to a crisis situation, you are not the cause even though the emotions may be directed at you. Sometimes just knowing this simple fact can allow you to interact with the individual in distress in more functional ways. Additionally, the realization that everyone sees the world in a different way may help you consider the source of the behavior before you react and become part of a conflict cycle.

Whether dealing with people in an emergency, inspection, supervisor/supervisee interaction, or departmental politics you should understand that everyone has a very different perception of reality. This can often mean the difference between a positive, successful resolution of a problem or an on-going cycle of conflict.

How might this help you in daily life? Let's use the scene of an annual building inspection as an example. As you arrive at the business of Mr. Jones, you have a chance to look over last year's inspection report. The report included several code violations and a comment about Jones' emotional reaction. As you get out of your car, you observe Jones walking toward you. At this point you have several options:

- You can expect Jones to act in a hostile manner and beat him to the punch by being "official" and nonresponsive.
- You can attempt to sidestep the issue by being overly friendly and conciliatory.
- You can realize that Jones is coming from a reality in which he may see "the inspector" as an enemy with whom he needs to be defensive. With this in mind you may surmise that Jones will be viewing himself as a victim of the system.

Obviously, just having good intentions may not change Jones' reaction to you. Should you begin to hear Jones denying the violations, shifting blame from himself to others, or directing anger at you, you have the choice of not reacting to his attempts at provoking you. You could open the discussion with Jones by remarking on how last years inspection seems to have not been a particularly positive experience for Jones. Tell him you want to work with him this year to "work out the kinks" so the inspection can be as positive and productive as possible.

## Taking Action To Avoid Conflict

Being aware of what is actually occurring allows you to create an environment for dealing with a difficult situation. Other direct actions you can take to minimize conflict are summarized below:

1. Attempt to see and hear the person — not just the problem.
2. Maintain an attitude of courtesy and respect for the person.
3. Reassure them of your desire to understand their perception of the issue.
4. Summarize what you hear the other person say before you respond to it.
5. Acknowledge what the person appears to be feeling by reflecting it to them.
6. Inform the person of what is occurring as it is going on. Lack of understanding is a major cause of conflict.
7. If strong emotion is being directed at you, acknowledge it and help redirect it toward the real target (e.g., the code rather than the inspector).
8. Be patient — remember that arguing is not helpful.

Although these suggestions are not meant to be taken as the definitive statement on conflict resolution, they may be helpful to you in developing your own unique ways of dealing with difficult situations. Remember to see the uniqueness in others and attempt to respond to that uniqueness with respect and compassion. If you accomplish this you will not be far from being able to successfully negotiate in conflict situations.

## SKILLS INVOLVED IN CONFRONTING ANOTHER PERSON

Many skills are involved in confrontation, such as communicating the observed behavior, your reaction to that behavior, and your interpretation of the situation. It also includes a desire to understand the other person's behavior. Understanding the behavior of another allows you to be more involved emotionally with that person, which minimizes his defensiveness when being confronted. Possible skills to use in confronting are:

1. Use statements that begin with a personal pronoun such as I, me, and my.

2. Use relationship statements, in which you express what you think or feel about the person with whom you are relating.

3. Use behavior description statements that clearly describe the visible behavior of the other person.

4. Describe your feelings using a personal statement. Put a name on what you are feeling and describe what action it makes you feel like taking. An example would be, "I am so confused that my head is spinning."

5. Give an understanding response by paraphrasing the statements made by the other person.

6. Give an interpretative response by **briefly** stating how you interpret the situation.

7. Do a perception check by stating what you perceive the other person's feelings to be.

8. Develop constructive feedback skills.

## CONSTRUCTIVE COMMUNICATIONS

Communication is valuable when it helps those involved examine their behavior and modify it in ways that improve their ability to understand others. The characteristics of constructive communication are:

1. **Empathy** — The communicator must understand the issues that are central to the other person's behavior and lifestyle.

2. **Timing** — The communication must be timed so that the person involved is open to receiving it without being overly defensive.

3. **Relatedness** — The communication must relate to the situation in which the two people are engaged. It should not occur unexpectedly.

4. **Concise** — The communication should be concisely stated and to the point. Long, ambiguous statements have an air of finality and tend to lose the receiver.

5. **Authenticity** — Communicate a genuine and sincere interest in the well-being of the person being confronted.

6. **Tentativeness** — Be slow to interpret the situation or to assume your interpretation is correct. An interpretation is a hypothesis about the other person's behavior, not a self-evident fact. Gather facts before interpreting.

---

# DO I FIGHT LIKE A SHARK OR REASON LIKE AN OWL?

Over the years I have observed-many styles of behavior that people use to resolve conflict. Some of the most common conflict styles remind me of animal characteristics. The following discussion compares some of the common conflict styles to certain animal behaviors.

**TURTLE** — Turtles withdraw into their shells to avoid conflicts. They give up their personal goals and relationships. They stay away from the issues over which the conflict is taking place and from the people they are in conflict with. Turtles believe it is hopeless to try to resolve conflicts. They feel helpless. They believe it is easier to withdraw, physically and psychologically, from a conflict than to face it.

**THE SHARK** — Sharks try to overpower opponents by forcing them to accept their solution to the conflict. Their goals are highly important to them and the relationship is of minor importance. They seek to achieve their goals at all costs. They are not concerned with the needs of other persons. They do not care if other persons like or accept them. Sharks assume that conflicts are settled by one person winning and one person losing. They want to be the winner. Winning gives sharks a sense of pride and achievement. Losing gives them a sense of weakness, inadequacy, and failure. They try to win by attacking, overpowering, overwhelming, and intimidating.

**THE TEDDY BEAR** — To teddy bears, the relationship is of great importance while their own goals are of little importance. Teddy bears want to be accepted and liked by other people. They think that conflict should be avoided in favor of harmony. They believe that conflicts cannot be discussed without damaging relationships. They are afraid that if the conflict continues someone will get hurt and that would ruin the relationship. They give up their goals to preserve the relationship. Teddy bears say, "I'll give up my goals and let you have what you want in order for you to like me." Teddy bears try to smooth over the conflict in fear of harming the relationship.

**THE FOX** — Foxes are moderately concerned with their own goals and about their relationships with other people. Foxes seek a compromise. They give up part of their goals and persuade the other person in a conflict to give up part of his goals. They seek a solution to conflicts where both sides gain something, the middle ground between two extreme positions. They are willing to sacrifice part of their goals and relationships in order to find agreement for the common good.

**THE OWL** — Owls highly value their own goals and relationships. They view conflicts as problems to be solved and seek a solution that achieves both their own goals and the goals of the other person in the conflict. Owls see conflicts as improving relationships by reducing tension between two people. They try to begin a discussion that identifies the conflict as a problem. By seeking solutions that satisfy both themselves and the other person, owls maintain the relationship. Owls are not satisfied until a solution is found that achieves their own goals and the other person's goals. They are not satisfied until the tensions and negative feelings have been fully resolved.

# SHORT COURSE IN HUMAN RELATIONS

The most important 6 words: "I admit I made a mistake."

The most important 5 words: "I am proud of you."

The most important 4 words: "What is your opinion?"

The most important 3 words: "If you please."

The most important 2 words: "Thank you."

The **most** important single words: "We."

The **least** important single word: "I."

**This page intentionally left blank.**

# POST-PROGRAM ACTIVITY 1

## SCENARIOS - MODULE IV

*Instructions:* The following are scenarios in which conflict is involved. If you were the fire officer, how would you handle each of these incidents?

### Scenario 1

As the on-duty fire officer, you are about to leave the station when you observe an off-going Captain remove a tool box from a pick-up truck and place the box in his car. Three shifts later you overhear the Captain who owned the truck complain that someone has stolen his tool box. Do you:

a. assume it is a practical joke and ignore it?

b. speak up and tell the Captain what you observed?

c. don't say anything right then, but talk to the Captain who moved the box and tell him what you observed?

d. get all of the Captains together and discuss the missing tool box?

Comments:_______________________________________________

_______________________________________________

_______________________________________________

_______________________________________________

_______________________________________________

## Scenario 2

A structure fire comes in about five minutes after shift change one morning. As a fire officer, you arrive on the scene of a smoke-charged structure. No fire is visible, but there is a lot of smoke. The Captain is sitting on the curb, head between his knees. He has vomited in the street. The engineer is running the fire. Upon closer observation, you can determine that the Captain has alcohol on his breath. He is a 30 year veteran, used to be your own supervisor, and he is currently experiencing a divorce. Do you:

a. order him to straighten up and get back to his job?

b. advise him he is suspended and order him to get back into the vehicle's seat and remain there until the fire is over?

c. ignore him and work with the engineer to get the fire out?

d. other (please specify)

In addition, what would you do upon return to the station?

Comments:_______________________________________________

_______________________________________________

_______________________________________________

_______________________________________________

_______________________________________________

_______________________________________________

## Scenario 3

3. The wife of one of your firefighters calls you. She advises you that her husband beats her up regularly and she wants you to warn him that if he doesn't stop, it can result in losing his job. Do you:

a. advise the wife that this is a personal matter and that you cannot get involved?

b. after getting the facts from the wife, give her counseling on where she can get additional professional help?

c. talk to the man, advise him you know of the problem, and warn him of the consequences?

d. other (please specify)

Comments:______________________________________________

______________________________________________

______________________________________________

______________________________________________

______________________________________________

## Scenario 4

Over a period of time, you have submitted numerous carefully researched suggestions and proposals you feel should be adopted since they would benefit the department. Some even involved no money expenditures. Unfortunately, 95% of the time your proposals are returned by your supervisor with a comment such as, "No, my philosophy does not agree with what you have suggested. As long as I am in charge here, we'll do it my way." You have reached the end of your rope and feel that your own morale will soon be effected. Do you:

a. give up on trying to help the department improve?

b. involve others in future suggestions so your superior has to fight all of you?

c. sit down and discuss with your supervisor and determine what his philosophy is with regard to change?

d. other (please specify)

Comments:_____________________________________________

_____________________________________________

_____________________________________________

_____________________________________________

_____________________________________________

# POST-PROGRAM ACTIVITY 2

## SCORING YOUR CONFLICT QUESTIONNAIRE

*Instructions:* From the Conflict Questionnaire that you completed as Pre-Program Activity 1 (page 99), transfer the rating you gave each proverb or statement to the appropriate blank below. Total the rating numbers that you listed in each column.

| TURTLE | SHARK | TEDDY BEAR | FOX | OWL |
|---|---|---|---|---|
| ____ 5. | ____ 3. | ____ 1. | ____ 4. | ____ 2. |
| ____10. | ____ 8 | ____ 6. | ____ 9. | ____ 7. |
| ____15. | ____13. | ____11. | ____14. | ____12. |
| ____20. | ____18. | ____16. | ____19. | ____17. |
| TOTAL | TOTAL | TOTAL | TOTAL | TOTAL |

A high score indicates a conflict style that you are comfortable using frequently. A low score indicates a conflict style that you would rarely use.

Review the five conflict styles (competing, avoiding, accommodating, collaborating, and compromise) listed on pages 109-112. List the conflict styles below in the order of frequency with which you use it. Begin with the conflict style you use most often.

1.)________________________________________________

2.)________________________________________________

3.)________________________________________________

4.)________________________________________________

5.)________________________________________________

# POST-PROGRAM ACTIVITY 3

## ARE YOU A CANDIDATE FOR BURNOUT?

The following items describe some common symptoms of burnout. Respond by circling the number that best reflects how you feel the question describes your feelings and attitude. Consult the Instructor's Guide (page 59) for a discussion of this activity. Use the following scale to determine which number to circle:

1 = Strongly Disagree

2 = Disagree

3 = Agree

4 = Strongly Agree

1. Do you find yourself frequently upset or irritable? 　　　1 2 3 4

2. Are you performing your job carelessly or mechanically? 　　　1 2 3 4

3. Do activities you once enjoyed no longer interest you? 　　　1 2 3 4

4. Are you withdrawing from key relationships in your life? 　　　1 2 3 4

5. Are you less communicative with close friends or loved ones? 　　　1 2 3 4

6. Have you overextended or overcommitted yourself in terms of time or energy? 　　　1 2 3 4

7. Are you tired of it all, feeling mentally or physically drained? 　　　1 2 3 4

8. Do you try to do everything equally well? 　　　1 2 3 4

9. Have you lost your sense of perspective, catastrophizing minor setbacks? 　　　1 2 3 4

10. Are you suffering more physical complaints: Headaches, insomnia, frequent colds, fatigue?                          1  2  3  4

11. Do you have a hostile or cynical attitude toward others?                          1  2  3  4

12. Do you have unrealistic standards of behavior or performance for yourself?                          1  2  3  4

13. Is your general feeling one of depression or sadness?                          1  2  3  4

14. Are you working harder but accomplishing less?                          1  2  3  4

15. Do you dread going to work in the morning?                          1  2  3  4

16. Do you find there isn't time for relaxation and recreation?                          1  2  3  4

17. Is your day filled with constant frustration and dissatisfaction?                          1  2  3  4

18. Do you feel you are inadequately compensated for the work you do?                          1  2  3  4

19. Are you unable to laugh at yourself?                          1  2  3  4

20. Do you feel you are more forgetful than usual? Example - appointments.                          1  2  3  4

(Copyright, 1981, Forbes Associates)

# **REFERENCES - MODULE IV**


Churchman, C. West. *The Systems Approach.* Dell Publishing, 1968.

Farber, Barry A. *Stress and Burnout in the Human Service Professions.* Pergamon Press, 1983.

Lewis David & Sharpe, Dr. Robert. *Thrive on Stress.* Warner, 1977.

Selye, Hans. *Stress Without Distress.* The New American Library, 1975.

**This page intentionally left blank.**

# MODULE V

## BUILDING YOUR TEAM

"Everything depends on people. Give persons a function
and responsibility, then hold them accountable,
and you'll be surprised with the results you'll get."
— Irving Shapiro

# MODULE V NOTES

# MODULE V

# BUILDING YOUR TEAM

---

## OBJECTIVES

To provide the viewer and participant with:

- a basic understanding of the concepts of motivating and demotivating personnel
- a basic understanding of the concepts of team building and participative management

---

## OUTLINE OF KEY POINTS

**Threats and Improvements —** In a modern setting individuals expect to receive fair treatment by the organization. The organization has a right to expect competent performance by the individual. Motivation should be a result of the work environment and not of fear or coercion. Maximum performance results from matching abilities of individuals with the needs of the organization. The fire service, as a paramilitary organization, operates using a structure that depends upon teamwork and personal motivation.

**Motivation Model —** The motivation model illustrates the relationship between the ability of a person and how motivation effects the degree to which they will use their abilities. Many of the factors that motivate individuals to achieve are a result of a supervisor's actions. Concurrently, most of the demotivation that occurs is a result of a supervisor's actions. No one can motivate people. People motivate themselves based on how they feel about the organization, their superiors, and if they feel threatened.

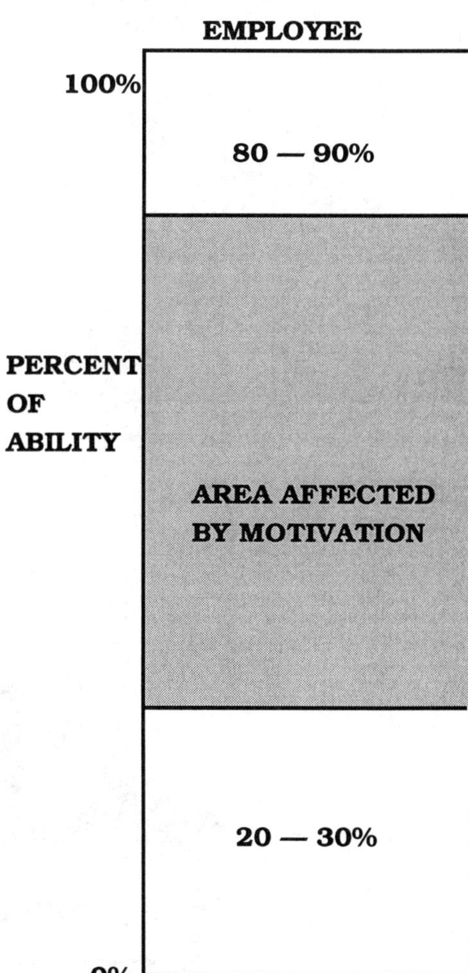


**Figure 5-1   Motivation Model**

**Motivation and Demotivation — *Maslow's Hierarchy of Needs*** identifies a series of levels of satisfaction that either motivate or demotivate a person to action. While many people have looked at the hierarchy as a steppingstone type approach that is continually escalating, it is actually a dynamic process that is undergoing change constantly. A person can function at different levels in the hierarchy depending upon the actions of their superiors and subordinates. An individual is unable to move to the next level until they have satisfied their needs at the level they are currently on. People won't have an adequate self-concept until they have worked through the lower levels and reached the esteem or self actualization level. No one can feel good about themselves when they are threatened. Each individual must be given an opportunity to perform at the highest level possible for them. No one will function at their highest possible level continually.

**Unmet Expectations** — No team can function unless there is agreement on what the team's goals and objectives are. Fire officers have expectations for their personnel. Subordinates have expectations of their superiors. Failure to achieve expectations by either party disrupts team effectiveness. The team must

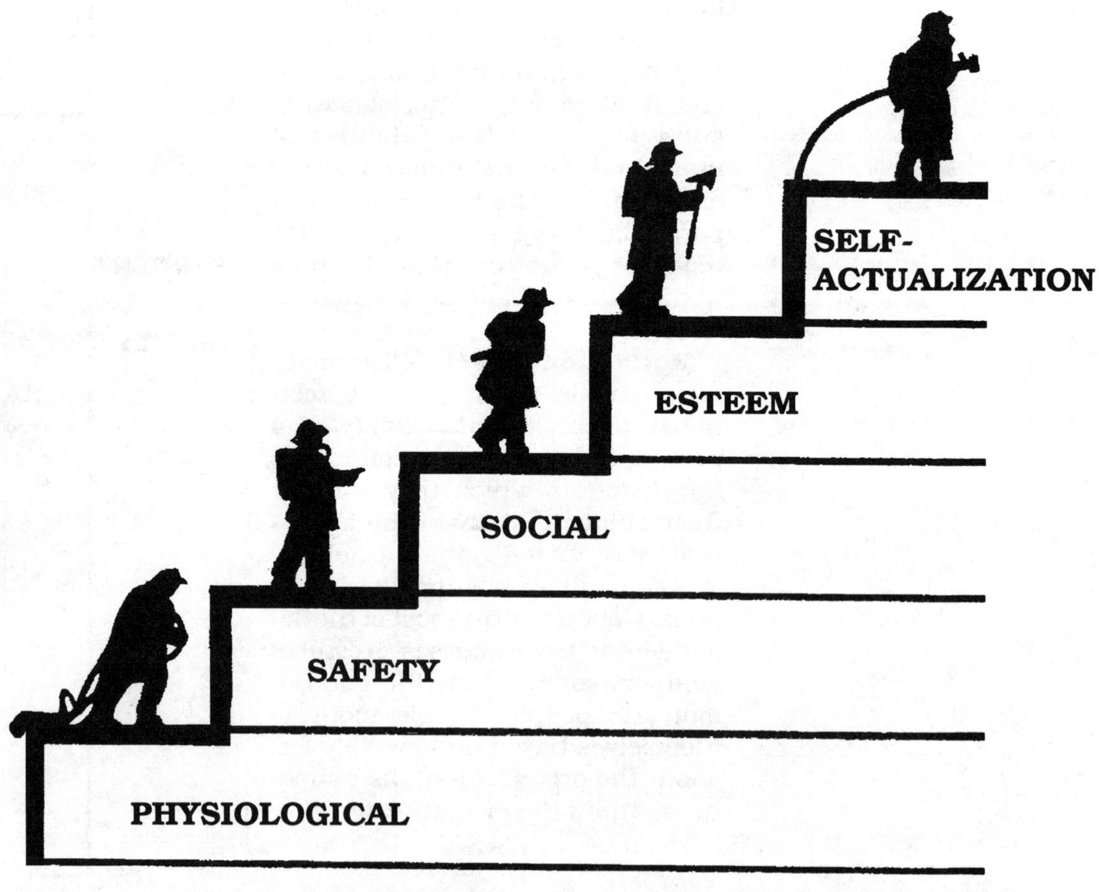


**Figure 5-2   Maslow's Hierarchy of Needs**

be goal oriented, and they must have a long-term reason for existing. Each fire company must set some objectives of how they can interface with the community on a long-term basis to cause the community to be a safer place.

**Loyalty** — J. Edgar Hoover once stated, "An ounce of loyalty is worth more than a pound of cleverness." Loyalty, in the fire service sense, does not mean blind obedience. True loyalty from a firefighter is loyalty to the goals that have been established for the organization, not loyalty to themselves as individuals working within the organization. Loyalty means preserving the integrity of an organization by supporting the organization in spite of differing opinions. However, there should be **organization honesty**, or a time when people within all levels of the organization are given the opportunity to express what they think and feel.

**One Minute Manager** — A popular theme today is on "One Minute Management." We are a 24 hour a day business. Teamwork is built a minute at a time, but is measured by the collective efforts of the fire company. This requires that an individual be a participant in the team and that the team gets progressively better or "grows." Morale in the firehouse is at its peak after a fire. This high level of morale is caused by the individuals in the organization taking what they do well and putting these skills to use. As a team, they were able to accomplish a common purpose. Working well as a team establishes morale, not just feeling good.

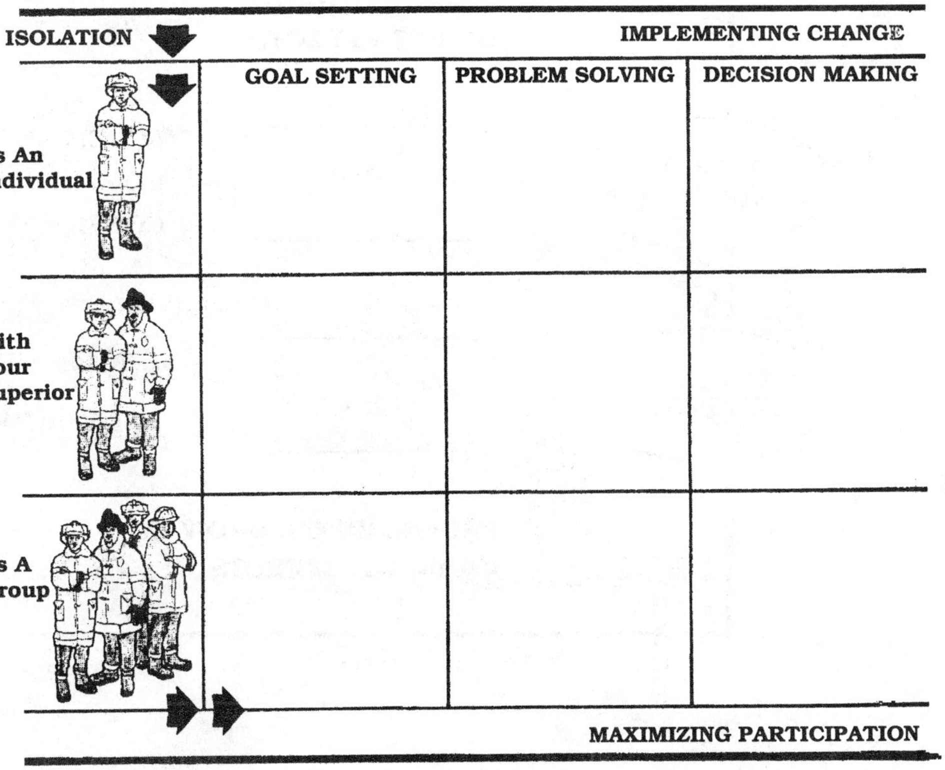

**Figure 5-3   Growth Model**

**Growth Model** — This model illustrates the relationship between the individual's participation in goal setting, problem solving, and decision making and a group's ability to effect change. The committment to the outcome heightens when participation in problem solving is allowed. Individuals feel they have gained a level of acceptance in the organization through participation. This allows them to move higher on Maslow's Hierarchy of Needs.

**Participation Model** — This model illustrates the relationship between the types and levels of participation and the improved performance of the individual and team. Individuals must be allowed to communicate how they believe they can best participate within the organization.

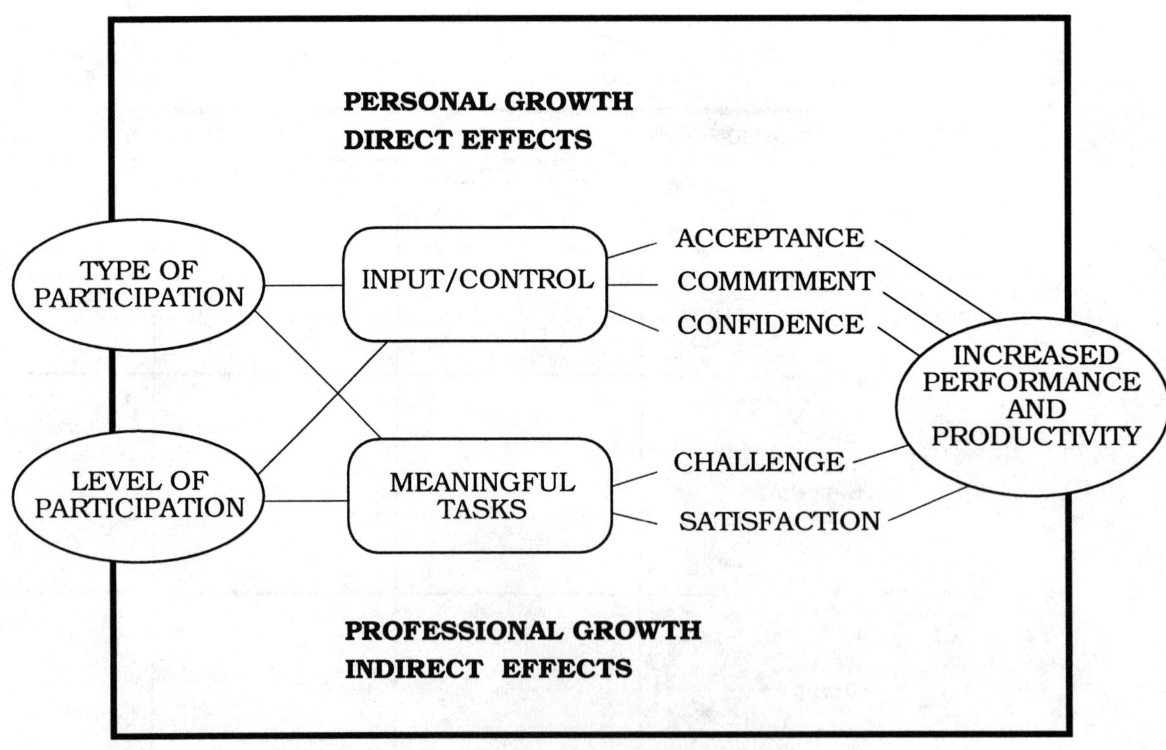


**Figure 5-4   Participation Model**

## WHAT IS A FIRE COMPANY?

A fire company is a group of individuals who must work both independently and interdependently in order to attain their individual and organizational goals.

1. The single, most important skill that fire officers have available to increase team effectiveness is their ability to make and maintain **effective** communication.
2. All the talent necessary to make the team/organization anything it wants to be is probably already present within the group.
3. The combined power and potential effectiveness for a team is greater than the sum of the contributing members.
4. Everyone already knows what to do and what they want. The focus, therefore, is on how the members assist or hinder themselves.
5. As a rule of thumb, more team dysfunction occurs through **conflict avoidance** than through conflict.
6. **Authentic** and realistic work relationships are much more productive than **"good"** working relationships.
7. Team members have to get along well enough to obtain the organizational objectives. It is all right to disagree, but not all right to be disagreeable.
8. The maximum potential for team strength and effectiveness is limited by the proportion each member fails to utilize his/her potential strengths.
9. The work, for the most part, has to be "fun."
10. Functioning as a team is a choice and not a natural occurrence of organizational structure. Individual and organizational objectives are best met through use of conscious team building efforts.

## WHAT THE EFFECTIVE FIRE COMPANY LOOKS LIKE

1. The company officer has high visibility.
2. The team is highly effective in all of its tasks.
3. The team is more concerned with doing the right things than merely doing things right.
4. Although not necessarily warm, relationships are fairly relaxed.
5. All members clearly understand and agree upon the team objectives.
6. Each team member knows how their **unique** contribution interfaces with the group's objective(s).
7. There is limited overlap between individual team members and their contributions to the team effort.
8. Team members are willing to voice agreement or disagreement openly, without fearing it might effect relationships.
9. The team's focus is constantly "RIGHT HERE, RIGHT NOW," but has the ability to foresee the future.
10. The team has the capability of maintaining flexibility and responsiveness to changing situations, i.e., to let go of procedures that are not working well.

11. The team has the capability of maintaining a "status quo" stance when appropriate.
12. Each team member is valued for their individual uniqueness and difference.
13. There is a norm that states, "You are free to be who and how you are. I am free not to like it." This norm must be used to contribute to team effectiveness, not in a negative way.
14. There is a norm that states, "it is OK for me to like some people more than others," so long as the open stated preferences do not result in discriminatory, unfair, or task destructive behaviors.
15. The team is mainly comprised of individuals who have complimentary talents and approaches rather than similar.
16. The team focuses on what it is doing that is effective AS MUCH as what is not going well.

---

## TEAM BUILDING

### Why A Team

Each manager has the potential for developing his work force into a team. The most important criteria for determining whether a "team" effort is needed is to answer the question, "Is there a need for team members to work together in order to meet organizational objectives, and if so, to what extent?"

There is no point in pursuing team development if the task or organizational objectives do not require a team effort. For example, a college faculty is generally made up of diverse talents. Each member is almost totally responsible for his "own thing" in terms of teaching, research, etc. There is little need for the faculty to interface with others in order to get the job done. In this example, the departmental chairman's function is one of coordinating the individual efforts rather than combining them into a team.

However, team development is very applicable at the managerial levels of an organization when the final output requires the efforts of several different departments or individuals. The question boils down to, "Can I achieve a higher output and individual satisfaction with an individual effort or through combining the efforts of several people?" The move toward team building clearly needs to be one of conscious choice on the part of management (the leader, supervisor, company officer, director, etc.).

On the following page, you will find a list of some conditions that would indicate a need for team building.

# Conditions Indicating The Need For Teambuilding

1. Loss of productivity.
2. Increase of grievances within the organization.
3. Evidence of improper hostility or conflicts.
4. Confused assignments, missed signals, or unclear relationships.
5. Apathy.
6. Work is done routinely, lacking imagination or initiative.
7. Ineffective use of personnel.
8. High dependency on or negative reactions towards supervisors.
9. Complaints from public about the quality of programs or services.
10. Continued increase in costs without an increase in productivity.
11. Difficulties in dealing with each other at a personal level.
12. Poor interaction with other fire companies or units.
13. Low commitment to team objectives.

## — OR —

1. Things are going well, but the team wants firmer control and awareness over the process.
2. A new crew or unit that needs to learn teamwork quickly.
3. Several functioning units are being merged into a new unit.
4. The company wants to diagnose the team's present behavior.
5. The organization needs restructuring because of a change in policy, technology, or command.

## TOTALLY INVOLVED OR OUT ON ARRIVAL?

It is interesting to examine how the jargon in a specific occupation developed. Every fire department has well-known expressions regarding the nature of emergencies. For example, the terms "loom-up," "header," and "all hands" are a form of verbal shorthand that tells the firefighters they have an emergency in progress. A frequently heard expression is "totally involved," or its counterpart, "out on arrival."

"Fully involved" means a working fire and "out on arrival" means nothing but taking the paperwork and returning to the station. These two terms may also be descriptive of the manner in which we and our organizations demonstrate our degree of professional involvement.

Are you "totally involved?" Recently I participated in a workshop with a group of young officers who questioned how a person was to become professionally competent if they were not allowed to become involved in the different levels of activities in the fire profession. Specifically, they were referring to the fact that the only person allowed to become a member of professional organizations outside of the department is the chief officer. Further, the only person who participates in meetings, committees, or activities is the fire chief. As a result, these young officers felt as if there was an obstruction to their ability to develop professionally.

As a chief officer it's O.K. to be "totally involved" yourself. That's one of the roles and functions of a leader or manager. You should be a role model for your organization by participating in organizations and being part of the decision making process in these outside professional organizations.

However, are your people fully involved? Do the subordinate members of your company participate in professional organizations, fire departments, or committees. Do they have to rely on you to bring back the information from meetings? If the answer to question one is "No" and the answer to question two is "Yes," then there is a possibility that you are working a lot harder than you should.

### Is Your Department A One-Man-Show?

*There should be a vertical structure within the staff that allows participation in a wide variety of organizations under a wide variety of circumstances.*

The chief officer should not be the "central switch" of all information flow in an organization. Instead, there should be a vertical structure within the staff that allows participation in a wide variety of organizations under a wide variety of circumstances. One of the most common complaints I hear from fire chiefs is that they must attend too many meetings. Careful analysis of that complaint reveals that they are probably attending meetings other people could just as easily attend. In some cases, other individuals could make a more significant contribution than the chief. This is not to say that the chief doesn't have a responsibility, or right, to monitor what is going on in different organizations. That is an obligation that cannot be shirked. What is being said is the fire chief needs to have combat intelligence coming from a wide variety of sources, instead of trying to be both scout and general at the same time.

One way you can analyze this phenomena is to look at the organizational structure of your department. On a piece of paper, place the names of your staff in a column.

Next to their name list the organizations they belong to and the regularly scheduled meetings they attend. If one individual seems overburdened with responsibilities and another is nearly free, then perhaps a redistribution of this workload should be made.

## Spreading The Information Around

*If one individual seems overburdened with responsibilities and another is nearly free, then perhaps a redistribution of this workload should be made.*

There is another element required to make this type of system function effectively, feedback and staff briefing. It is extremely expensive to maintain multiple memberships in organizations just to be able to read the newsletters and bulletins distributed by these affiliations. Instead, it is more effective to develop a routing slip that can be attached to the organization's literature and pass it among the staff. That way, the extra cost of duplicating materials can be avoided. At the same time, you have documented proof that the individuals on the routing slip have had an opportunity to review the material. Requiring the person's initial on the routing slip as it crosses his/her desk assures their momentary interest in the material.

The material should then be collected in a central repository for future reference. When visiting fire stations, you often find huge stacks of old issues of fire magazines stuck away in closets and desk drawers.

A few moments spent organizing those materials may provide an excellent resource for department training officers or company officers looking for material to add to a drill.

It should be clearly established who will attend what meetings. Then staff briefings should include an update regarding issues and projects underway with the different organizations. Individuals should report on the activities they have attended. This results in sharing large amounts of information in a short period of time.

It should also be a requirement that after personnel attend a departmental sponsored training activity they provide a written review of the material covered. This feedback can be important in determining whether other personnel will be sent to the classes. In addition, it may provide the training officer with a new resource for training material that can be used in the department.

## Greater Involvement Will Create Change

In discussing this concept with fire officers, I discovered some individuals who are extremely reluctant to allow their personnel to become more "fully involved." One chief officer indicated that one of his biggest problems is individuals returning from school with new ideas that he has to field in staff meetings. From that individual's point of view, more involvement on the part of his subordinates merely meant more pressure upon him to

make decisions. The fact is, that is true.

The more involved individuals become, the more likely they are to place pressure on the hierarchy to become responsive to the need for change. In some organizations, placing restraints on individuals' participation in group activities may be an effective way of limiting change in an organization. On the other hand, nothing is as powerful as an idea whose time has arrived. Organizations that remain unaware of the growth and development that occurs from trying new ideas, soon find themselves obsolete and politically unstable.

To continue with the analogy of a structure being fully involved, it is also apparent that it doesn't happen instantaneously. The development of participation in an organization goes through definite cycles, like a fire that starts with a point of origin and goes through specific stages of growth. These cycles form a ring-like relationship among the following elements:

- a problem which creates...
- searching for new answers that allows for an...
- opportunity to gather knowledge followed by the...
- implementation of new concepts providing the groundwork for...
- evaluation of successes and failures, repeated by...
- new problems.

Organizations that realistically appraise their problems cannot help but become more involved. The number of people involved in the problem solving process is usually relatively small at the outset. As an individual's participation results in a more effective work environment, more people become

involved in the process and a synergistic effect begins to take place.

It is interesting to note that often there is a high degree of involvement at the bottom with respect to the firefighter's associations and unions, but a limited degree of involvement at the upper staff levels. Conversely, there are many other organizations that languish at the lower levels, in spite of a high degree of professional involvement by only one or two individuals at the upper levels. The best situation is when the staff is totally involved in managing an organization and the personnel is highly involved in professional development and program management.

Who knows? If this idea of participative management really works and our personnel become fully involved, perhaps it will result in an upheaval in modern fire management. A totally involved fire department will not only improve the profession, but will improve the quality of life in the community it is responsible to protect. The best place to start is yourself. Whatever position you hold — firefighter, company officer, battalion chief, or chief — getting your people fully involved is the first step in building your team.

*The more involved individuals become, the more likely they are to place pressure on the hierarchy to become responsive to the need for change.*

# GOAL SETTING — THE KEY TO SUCCESS

One of the best ways to get your crew or department "fully involved," is through the use of goal setting. What is a goal? A goal is a target, something to shoot for. A goal is something to be achieved at a specific time in the future. Goal setting for the organization, employee, officer, and fire company can be a key tool for success. Having a specific target gives specific direction. All of us like to feel that we are accomplishing something. When we have individual, company, and organizational goals, then we know we are making specific progress.

If you want to build your team, goal setting is an excellent place to start. In Module I we talked about the Pareto Principle, which states that eighty percent of our events and activities result from the performance of twenty percent of the population. Goal setting can be a terrific way to focus in on the needs of your people, and open lines of communication which had not been opened before.

Goal setting also benefits the organization as it is able to maintain or increase its effectiveness. Employes also benefit from goal setting. When goals are achieved, employees can earn rewards that are meaningful to them.

The fire officer benefits as much as anyone from successful goal setting. In setting goals for yourself, assisting your firefighters in goal setting, or setting company goals, you can clearly demonstrate what has been accomplished. Goal setting allows you to clearly show the achievements to your firefighters, supervisor, and yourself.

## Goal Setting Involves Three Basic Steps

1. **Fix a Target** — Fixing a target means setting a specific goal. Spell out exactly what will happen at a particular time. Use words which identify action such as, complete, list, etc. A goal is relevant if it addresses an activity that makes a difference in overall performance.

2. **Plan How to Reach the Goal** — Outline the specific steps needed to reach the goal. This may involve physical changes, such as rearranging the medical kit, or teaching a new skill. Whatever changes are necessary, they should relate to reaching the goal. For goals to be attainable they need to be practical, reasonable, and achievable in the time specified for them.

3. **Measuring Progress** — In order to measure success you need to put a tracking system in place. There must be some plan for measuring progress. By checking progress periodically and making the necessary changes, you can assure the goal is reached.

## Three Types Of Goals:

### Long-Range Goals

Many organizations, departments, and individuals set long-range goals. These are goals that stretch into the future for three to five years, or even longer. Long-range goals are often those that concern growth, development, or expansion.

### Intermediate Goals

Often organizations, departments, and individuals set goals for an entire year or more. Sometimes an annual goal is broken down still further. Semiannual or quarterly goals should serve mainly as steppingstones toward annual goals. They also provide a way of measuring progress toward annual and long-range goals.

### Short-Term Goals

Short-term or temporary goals are often task goals related to a specific job. For instance, introducing a new form or procedure may have a time period of one week or even one day. These are short-term task goals. Sometimes a new task will arise unexpectedly. Then you and your employees will have new goals to set. For instance, let's say that you suddenly have a special report to get out. You and your employees may have to work together to establish goals for each person so the overall goal of completing the report in a short period of time can be met.

You should remember the basic rules of goal setting are:

- fixing a specific target
- planning on how to reach the goal
- and planning how to measure progress toward the goal.

At the end of this module you will find an individual goal worksheet and a company action plan. You are encouraged to fill out a personal action plan, assist in developing a plan for each of your personnel, and hold a company meeting to establish a company action plan for a given period of time.

Goal setting can be the first step in motivating your personnel to become more involved, building positive lines of communication, and providing a cornerstone for your team building program.

## WHAT IT TAKES TO BE NO. 1

"Winning is not a sometime thing — it's an all-the-time thing. You don't win once in a while, you don't do things right once in a while, you do them right all the time. Winning is a habit. Unfortunately, so is losing.

"There is no room for second place. There is only one place in my game and that is first place. I have finished second twice in my time at Green Bay and I don't ever want to finish second again. There is a second place bowl game, but it is a game for losers played by losers. It is and always has been an American zeal to be first in anything we do and to win and to win and to win.

"Every time a football player goes out to ply his trade he's got to play from the ground up — from the soles of his feet right up to his head. Every inch of him has to play. Some guys play with their heads. That's O.K. You've got to be smart to be No. 1 in any business. But more important, you've got to play with your heart — with every fiber of your body. If you're lucky enough to find a guy with a lot of head and a lot of heart, he's never going to come off the field second.

"Running a football team is no different from running any other kind of organization — an army, a political party, a business. The principles are the same. The object is to win and to beat the other guy. Maybe that sounds hard or cruel. I don't think it is.

"It's a reality of life that men are competitive and the most competitive games draw the most competitive men. That's why they're there — to compete. They know the rules and the objectives when they get in the game. The objective is to win — fairly, squarely, decently, by the rules — but to win.

"And in truth, I've never known a man worth his salt who in the long run, deep down in his heart, didn't appreciate the grind, the discipline. There is something in good men that really yearns for, needs, and desires the harsh reality of head-to-head combat.

"I don't say these things because I believe in the 'brute' nature of man or that men must be brutalized to be combative. I believe in God, and I believe in human decency. But I firmly believe that any man's finest hour — his greatest fulfillment to all he holds dear — is that moment when he has worked his heart out in a good cause and lies exhausted on the field of battle — victorious."

**- Vince Lombardi**

# MASLOW'S HIERARCHY OF NEEDS

| NEEDS | EXPRESSION | THREAT | ACTION |
|---|---|---|---|
| **Self-Fulfillment** | Realizing potential<br>Creativity<br>Self-development<br>of talents<br>Self-projection<br>of opportunities<br>Individualism | Inflexible policies<br>Narrow definitions<br>of job tasks<br>No interest in job<br>enrichment | Trust<br>Progressive Climate<br><br>Tolerating failure<br>Management<br>Development programs<br>Theory V style |
| **Ego Needs** | Self-esteem mastery<br>(confidence)<br>(self-confidence)<br>(independence)<br>(achievement)<br>Reputation<br>(recognition)<br>(appreciation)<br>(respect)<br>(personal identity) | Job fragmentation<br>Careless criticism<br>Failure to praise<br>or reward<br>Centralization of<br>authority<br>Accounterments of<br>office<br>Theory X style | Decision consultation<br>Use of rewards &<br>recognition<br>Job enlargement<br>& influence<br>Subordinate feedback<br>Visibility &<br>exposure<br>Use of goals |
| **Social Needs** | Belonging<br>Association<br>Acceptance<br>Giving & receiving<br>attention<br>Friendship<br>Love and to be<br>loved | Unfriendly assoc-<br>iates &<br>management<br>Office cliques<br>& rejection<br>Poor Communication<br>Ostracism | Participation in<br>management<br>Morale & opinion<br>Survey<br>Communications &<br>special bulletins<br>Project management<br>Theory Y style |
| **Safety Needs** | Protection against<br>a. danger<br>(arbitrariness)<br>b. threat<br>c. deprivation<br>d. insecurity<br>Unpredictability<br>in environment | Arbitrary action<br>by management<br>Sudden changes<br>Poor communication<br>Lack of policy &<br>procedure<br>Favoritism &<br>meetings<br>Discrimination<br>No benefits<br>Theory U style | Performance appraisal<br>system<br>Written policies<br>& procedures<br>Position guide<br>Newsletters<br>Bulletin boards<br>Grievance process |
| **Physiological Needs** | Air, food, water,<br>space, rest &<br>exercise<br>Protection from<br>the elements | Inadequate benefits<br>Poor pay<br>No break<br>Cramped conditions<br>Poor facilities<br>Excessive hours | Sound & effective<br>safety program<br>Good working<br>environment<br>Work planning &<br>scheduling<br>Rest, lunch, &<br>holiday periods<br>Shorter work week |

# POST-PROGRAM ACTIVITY 1

## TEAM ASSESSMENT

*Instructions:* The following checklist is an example of a team assessment form. Complete it for yourself as a member of the group. After completing the checklist, compare your perceptions with other members of the group.

| CHECKLIST | YES | NO |
|---|---|---|
| Do we have an informal, relaxed atmosphere? | ____ | ____ |
| Is there a sense of purpose to the group? | ____ | ____ |
| Does the team understand and agree upon priorities? | ____ | ____ |
| Does joint participation influence our decision making? | ____ | ____ |
| Does a person feel comfortable disagreeing with others? | ____ | ____ |
| Does the group have a standard of performance? | ____ | ____ |
| Does the group look for consensus when appropriate? | ____ | ____ |
| Is there group support and cohesiveness? | ____ | ____ |
| Has team performance improved in the last year? | ____ | ____ |
| Do all concepts and ideas get equal consideration? | ____ | ____ |

# POST-PROGRAM ACTIVITY 2

## INDIVIDUAL GOALS

**Goal #1**___________________________________________________

___________________________________________________

How is this goal to be achieved?___________________________

___________________________________________________

How and when will progress be measured?___________________

___________________________________________________

Expected date for accomplishing this goal:__________________

**Goal #2**___________________________________________________

___________________________________________________

How is this goal to be achieved?___________________________

___________________________________________________

How and when will progress be measured?___________________

___________________________________________________

Expected date for accomplishing this goal:__________________

**Goal #3**___________________________________________________

___________________________________________________

How is this goal to be achieved?___________________________

___________________________________________________

How and when will progress be measured?___________________

___________________________________________________

Expected date for accomplishing this goal:__________________

# POST-PROGRAM ACTIVITY 3

## COMPANY ACTION PLAN

**Goal #1**_______________________________________________

_________________________________________________________

How is this goal to be achieved?_________________________________

_________________________________________________________

How and when will progress be measured?___________________________

_________________________________________________________

Expected date for accomplishing this goal:__________________________

**Goal #2**_______________________________________________

_________________________________________________________

How is this goal to be achieved?_________________________________

_________________________________________________________

How and when will progress be measured?___________________________

_________________________________________________________

Expected date for accomplishing this goal:__________________________

**Goal #3**_______________________________________________

_________________________________________________________

How is this goal to be achieved?_________________________________

_________________________________________________________

How and when will progress be measured?___________________________

_________________________________________________________

Expected date for accomplishing this goal:__________________________

# POST-PROGRAM ACTIVITY 4

**SCENARIOS - MODULE V**

**Scenario 1**

**Setting:** Firefighter talking with Company Officer in office.

**Firefighter:** Got a minute?

**Company Officer:** Of course, Jack. Come on in.

**Company Officer:** What's up?

**Firefighter:** Well, you've probably noticed that I've had an attitude problem lately. I just can't seem to get excited about the job anymore. When I was new, I was real excited, but lately I don't know what's wrong with me. It's not you — I like our crew and all. I guess I'm just not as revved as I used to be.

**Company Officer:** I am aware that you haven't seemed happy lately. Sounds like you are feeling bored at work. Do you think that's an accurate description of how you're feeling?

**Firefighter:** Yeah, I guess I am bored. I don't look forward to coming to work anymore. Maybe I ought to quit.

**Company Officer:** You're a good, solid firefighter, Jack. You don't need to quit to get some energy back into your life. What about this job has interested you most in the past?

**Firefighter:** Well, there are several areas I like. I like the training we do with other crews. I like the inspections, but I don't feel real confident doing them because I don't know as much as you do about the codes. Of course I like the fires, it's just that we don't get enough of them.

**Company Officer:** You know, I've been watching you on inspections. I think you handle people very well. You explain the codes to business people as well as anyone. Of course there is always room for being more familiar with the codes, but I think you have a good basic grasp of the intent. That's more than I can say for most firefighters!

**Firefighter:** Yeah, I think we provide an important service to the business people during inspections. I try to read up on the codes.

**Company Officer:** Would you be willing to be in charge of our company's inspections? I mean the whole thing — scheduling, paperwork, follow-up — the whole shebang?

**Firefighter:** I don't know if I could handle it...

**Company Officer:** Sure you can! I'll still be around to assist and train you in the paperwork, that's no problem. I think I can even get some extra training for you. There's a fire code class coming up next month. Well, what do you think?

**Firefighter:** That would be great! I do have some ideas about scheduling our inspections that might help our workload.

**Company Officer:** Well Jack, I do have a selfish motive as well. If you head up the inspections, I would be freed up to spend more time on some of my other duties.

**Firefighter:** Wow! This is great! Thanks for the vote of confidence. I won't let you down.

**Company Officer:** I'm sure you won't let us down, Jack. And, thank YOU!

---

Evaluate this scenario and describe the positive and/or negative content. If you were the company officer, would you have done anything differently?

_______________________________________________________________

_______________________________________________________________

_______________________________________________________________

## Scenario 2

**Setting:** Engine company about to get on the engine and leave, when the chief stops them.

**Battalion Chief:** Hey Bob! Wait!

**Company Officer:** What Chief?

**Battalion Chief:** I need for you to do something for me. Have you got time?

**Company Officer:** Well, we were just leaving to do some inspections. I made an appointment with that body shop I've been trying to get in for awhile.

**Battalion Chief:** It can wait. Downtown just called. They want to know how many times other engines responded into Station 1's area last year. I have to go out for awhile, so I was wondering if you could get those numbers for me.

**Company Officer:** But Chief, I made an appointment with this guy. It's really important to me th—-

**Battalion Chief** (interrupting): And it's important to me that you do this job! Weren't you just telling me the other day that you wanted to take on more responsibility? Now when I ask for your help you give me a hard time!

**Company Officer:** Can't downtown wait? I can do that for you as soon as I get back.

**Battalion Chief:** LATER?! Am I to understand that you are refusing to follow a reasonable REQUEST from your supervisor?

**Company Chief:** No, sir. We'll get on it right away. I'll just call and cancel the inspection with the body shop.

---

Evaluate this scenario and describe the positive and/or negative content. If you were the company officer, would you have done anything differently?

________________________________________

________________________________________

________________________________________

## Scenario 3

**Setting:** An officer new on the shift is briefed by the battalion chief.

**Battalion Chief:** I just wanted you to know, Bob, that I'm not one of those new, touchy-feely supervisors.

**Company Officer:** Uh Huh?

**Battalion Chief:** I have worked my way up through the ranks to get where I am. I've seen just about every emergency there is to be seen. I guess that I'm just not comfortable with the emphasis they place on communication stuff now.

**Company Officer:** I'll try to remember that.

**Battalion Chief:** In other words, it's not likely that I will come by and tell you when you've done a good job. I expect my officers to do a good job, so why should I have to tell them? But you can count on hearing from me when you screw up!

**Company Officer:** Okay, so no news means good news, right?

**Battalion Chief:** Right. I want you to know that I listen on the radio to all the calls that happen in my area, even if I'm not there. If I think you did something wrong I'll let you know. I'm particularly strict on safety.

**Company Officer:** So am I, Chief. I always stress safety during training, as well as on scenes. You shouldn't have any problems with my performance in the area of safety.

**Battalion Chief:** Well, I'd better not. I also rely on the more experienced officers to tell me how they think you are doing on the scenes that I can't get to. Well, that's about it from me. Welcome to the shift. Do you have any questions?

––––––––––––––––––––

Evaluate this scenario and describe the positive and/or negative content. If you were the company officer, would you have done anything differently?

––––––––––––––––––––––––––––––––––––––––––––

––––––––––––––––––––––––––––––––––––––––––––

––––––––––––––––––––––––––––––––––––––––––––

# REFERENCES — MODULE V

Gellerman, Saul W. *Management by Motivation.* American Management Association, Inc., 1968.

*Fire Service Supervision: Increasing Team Effectiveness.* National Fire Academy, National Emergency Training Center.

# MODULE VI

## DEVELOPING YOUR GAME PLAN:

## GETTING THE JOB DONE

"We trained hard... but every time we approached excellence, we were reorganized. I learned later in life that we tend to meet any new situation by reorganizing; and a wonderful method it can be for creating the illusion of progress while producing confusion, inefficiency, and demoralization."
— Petronius Arbiter, Roman Officer, 76 A.D.

# MODULE VI NOTES

# MODULE VI

# DEVELOPING YOUR GAME PLAN: GETTING THE JOB DONE

---

## OBJECTIVES

To provide the viewer and participant with:

- a basic understanding of the process of time management

- a basic understanding of their scope of responsibility in achieving the goals and objectives of the organization.

---

## OUTLINE OF KEY POINTS

**Overview** — The term "game plan" comes from the athletic field, but it can be applied in the fire house. If we don't have a sense of direction or keep score, there is no way to measure our contribution as individuals or as an organization.

**Hierarchy of Responsibility Model** — This model illustrates the relationship between the direction of an organization and the ability to produce results (see Figure 6-1). What is accomplished at the bottom of the organization is a direct result of the decisions made at the top of the organization. The following are components of the Hierarchy of Responsibility that will determine the direction of the organization:

- **Mission** — The mission clarifies the purpose of the organization and places it in context with other organizations with parallel missions. The mission should be written out in the form of a **Mission Statement.** This statement should be publicized and displayed for the benefit of all members of the organization.

- **Policies** — Policies clarify the pervasive philosophy of the organization and establish the ethical considerations of the organization.

- **Programs** — The specific areas of activity that have been chosen by the organization to achieve its mission. Your programs will be a reflection of your mission statement. If the mission statement is changed, then programs will change. Programs must have goals if they are to be useful to the organization.

- **Procedures** — The standards of performance or standard operating procedures establish minimum requirements in the organization. This involves time management.

- **Priorities** — The relative importance of an activity or resource in relation to achieving the goals and objectives.

- **Daily Practices** — These are the everyday methods

## DIRECTION OF THE ORGANIZATION

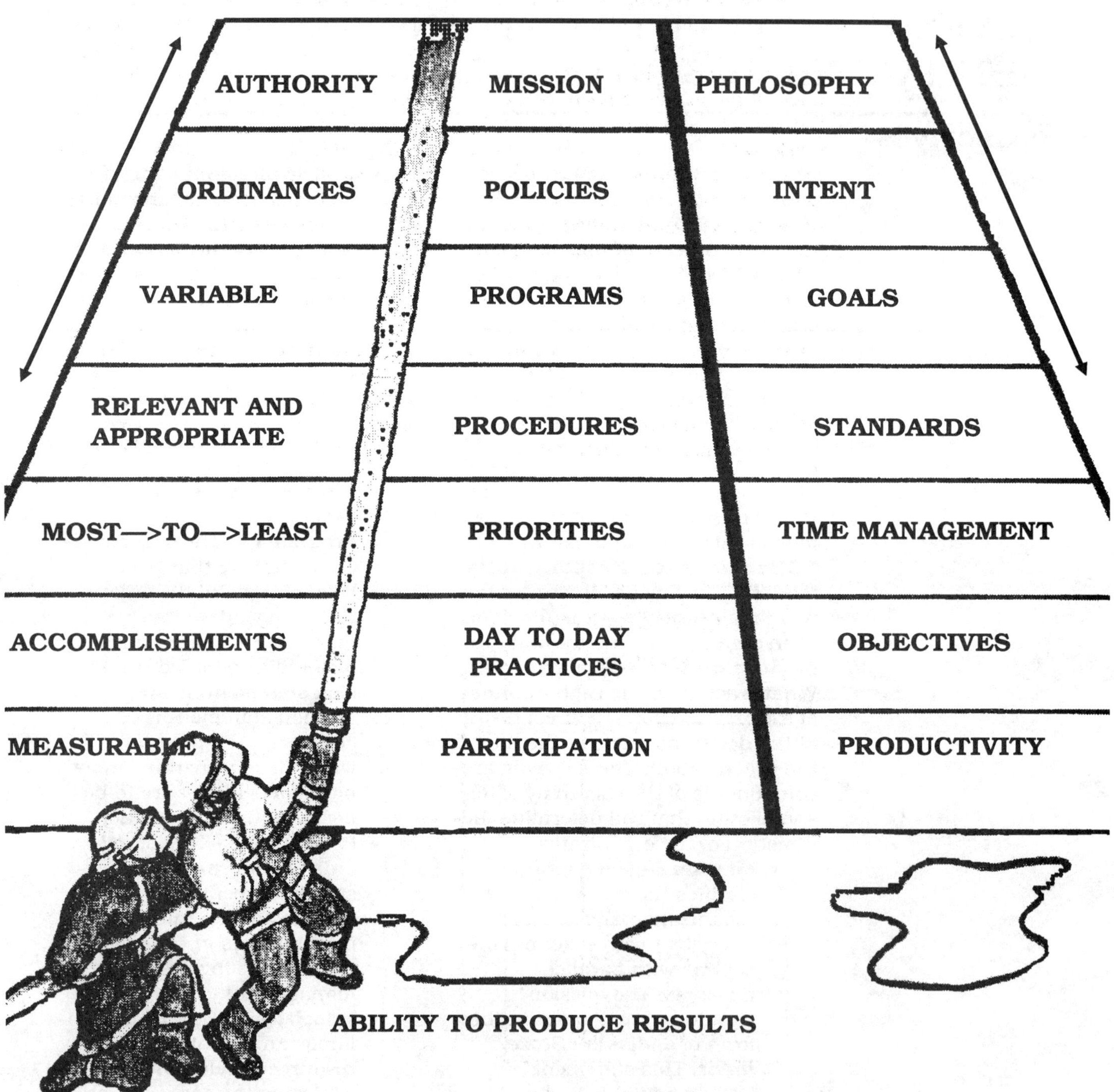


Figure 6-1   Hierarchy of Responsibility

used to support program activity.

- **Participation** — The actual achievement of tasks on an everyday basis.

**Goals and Focus** — The existence of goals and objectives by an individual, a fire company, shift, work unit, or entire department provides focus and visibility to its achievements. An organization that does not have identified goals is just existing without real direction.

**Goal Model** — This model illustrates the relationship between what a person wants to achieve and what the organization wants to achieve. The outcomes of a goal model can range from destructive conflict, to negotiated agreement, to achievement of difficult goals.

**Champions or Amateurs** — The greatest difference between amateurs and professionals is their level of acceptance of personal strengths and weaknesses. Acceptance that others they compete with possess greater skill than their own leads to growth.

**Objectives** — An objective is an agreement to achieve certain results during a specified period of time. Most organizations have vague and generally stated objectives. Clearly defined objectives have specific measurable milestones.

**Productivity Model** — Productivity is defined as the relationship between the quantity of goods and services produced and the amount of resources required to produce

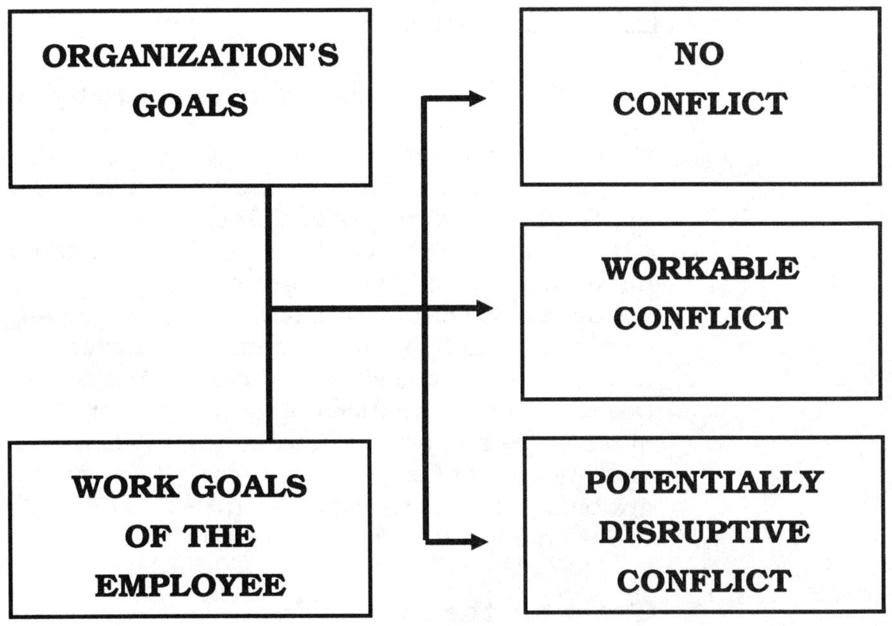


**Figure 6-2   Goal Model**

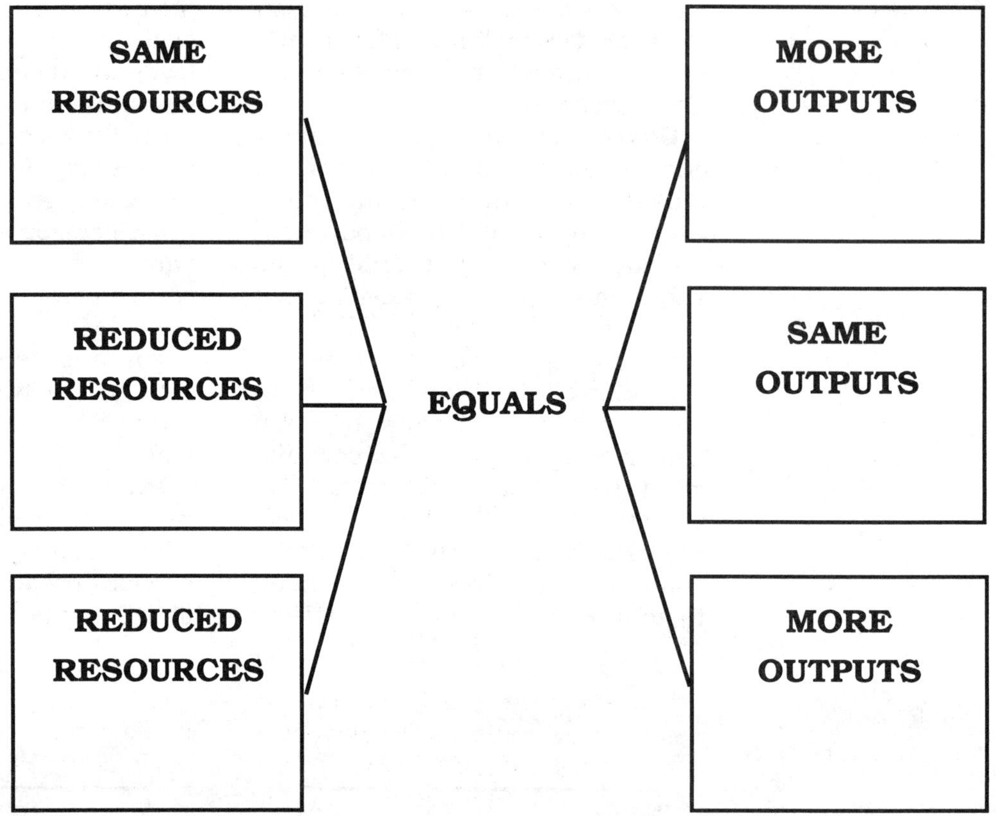


**Figure 6-3  Productivity Model**

them. A comparison of productivity over time or among different activities is stated as a ratio of output to input. In the fire service inputs are manpower, supplies, programs, and facilities. Outputs are service, citizen satisfaction, and community facilities produced. If you can take the same or reduced amount of resources and produce the same or higher level of output, then you are being productive. Some of the major changes the fire service will deal with in the future relate to this model. Many departments experience budget cuts, but are expected to maintain the same level of output.

**Keeping Score of Winners and Losers** — How can you tell if you are winning the game if you don't keep score? How can you determine whether you are a professional or amateur unless you compare your score against a set of standards.

## Tools for the Fire Officer

**Monthly Planning vs. Day-to-Day Crisis** — The use of monthly planning calendars and time management techniques establishes credibility for the fire officer. Fire axes are used to ventilate a burning structure. The fire officer uses different tools behind the desk. These

154

tools allow for a game plan to be established. These tools let you know where you are headed and what you need to accomplish in a given period of time.

**Responsibility** — The fire officer must take responsibility for everything that occurs in the fire station.

**Accountability** — The responsibility to manage resources to achieve the objectives of the fire company rests on the company officer. Tasks can be delegated. Accountability cannot be delegated.

**Calendars** — Calendars are time management tools and should be completed well in advance.

## Utilization of Shift Time

**First Ten Minutes of the Day** — The period of time spent in planning the rest of the day is very important. The first ten minutes of each shift or work day should be devoted to identifying that day's priorities. Incremental planning is a crisis orientation that damages the group's productivity and the officer's credibility.

**Time Management Concepts** — An individual should develop their own version of time management techniques. The reference section at the end of this module identifies several excellent resources.

**Monkey Management** — Tasks that have been delegated to others to achieve should not return to haunt you. Only delegate to those who are capable of completing the requested task, then hold them accountable to complete it correctly and in a timely fashion. William Oncken, Jr., in his book *Managing Management Time*, suggests that a good leader or manager should never carry more monkeys on their back than are absolutely necessary.

## Volunteer Departments

**Do Volunteer Departments Have to Manage Their Time?** — If an organization has a mission, sets goals and objectives, establishes priorities, and has tasks to be performed, it is essential they use time management. Developing a good game plan has nothing to do with whether you are being paid to do a job.

## Teamwork and Productivity

**I vs. We** — The fewer people involved in achieving an objective, the longer it will take. Participation and productivity are mutually supportive. Remember that "we" need to accomplish this goal, not "I" need to.

**Us vs. They** — The more conflict there is between work units, the longer it takes to achieve goals. Communication and cooperation are mutually supportive. Us vs. They is counter productive. The officer needs to use the abilities of those in his department and abilities available in other departments to provide an "us" atmosphere.

# TIME MANAGEMENT



I love a sign one of my friends has in his office. On his desk is one of those little cardboard signs you see in gift shops. It's tattered around the edges, it has obviously been moved from office to office. It states, "If you want to kill time around here — try working it to death!"

Time management is probably one of the most critical problems facing the fire officer. All of the crisis and conflict that occurs on a daily basis causes considerable competition for our time. There will never be more working hours in the day, so it is important for the fire officer to make maximum use of the time that is available.

There are several excellent books available on the subject of time management. One of the best of these is written by Alex MacKenzie entitled *The Time Trap*. This book has been published by McGraw-Hill Paperbacks. Two other books on this subject are *Time: How to Make the Most of Your Life*, authored by Alan Lakein, published by Signet, and *Getting Things Done*, authored by Edwin Bliss, a Bantam book.

All of these books suggest the same approach to time management — have a system and use it! They offer a wide range of suggestions on how to prioritize your time, set up systems that allow you to maximize your time, and so on. The suggestions are excellent, but most of them leave out one of the most important aspects of good time management — attitude.

If I were to ask, "Are you important?", I doubt very seriously that you would give me a negative answer. The next question is, "Is your time important?" Generally speaking, if the answer to the first question is "yes," the answer to the second question should be "yes" also. It is amazing how many important people allow their time to be either eroded or compromised. We allow others to steal one of our most precious possessions — our time.

There are several skills that you can develop to help you maintain control of your time. The use of these particular skills will increase your output. They will enhance your impact on your organization, city, and profession. Failure to develop these skills can have the opposite effect.

## Developing Time Management Skills



The first ability to develop is the art of **"developing your game plan."** For the fire officer, developing your game plan means deciding what has to be done, when, by whom, and how. It also means scheduling time for yourself and/or your firefighters to accomplish what has been planned. The concept of team planning can work well for the line officer. It provides direction for the company schedule and motivates your firefighters to participate in getting the job done.

The second ability to develop is **"expect the unexpected."** One of the biggest challenges to the fire officer is balancing the demand for our emergency services and the pressure we face to accomplish our routine duties. When developing your game plan allow time to achieve the things you have set out to do, as well as time for unexpected

156

events. One of the most frustrating things for you and your subordinates, is to develop a plan that has little chance of succeeding. Anticipate emergency response interruptions. Creating a game plan that has flexibility, while meeting the goals you and your subordinates have set, will reduce everyone's stress.

The third ability to develop is **"the art of delegation."** I have heard many people say they can't delegate, they have to do everything themselves. Yet, the ability to delegate can double or triple your output. When delegating, you must give your subordinates sufficient knowledge of the goals and objectives to be achieved for them to understand your intent. It is unlikely that any job completed through delegation will be perfect. Therefore, delegation is best used in circumstances where there is an opportunity for subordinates to learn, but a margin of error is acceptable.

*The ability to delegate can double or triple your output.*

## Managing Your Telephone Skills

The fourth ability is to develop a **"telephone protocol."** The telephone is basically a rude instrument. If people would have realized how much others' priorities would impact us through the use of the telephone, I doubt if Alexander Graham Bell would have been as successful in getting his invention accepted. No matter what is happening in your office or what priorities are, when the telephone rings we usually answer it. You are then operating on someone else's time, not your own.

There are three different techniques you can develop to help control the element of time with respect to telephones. The first of these is to ***make a conscious decision not to answer the telephone at certain times.*** A person was once sitting in my office waiting for me to finish a project when the phone rang. After letting the phone ring about six times, my friend picked up the phone and answered it for me. I asked him to tell the caller I would return the call. Later he asked why I didn't answer the telephone. I explained that I didn't want to break my concentration on the project I was involved in to respond to someone else's priority.

Another technique you can learn regarding telephone protocol is to learn to ***get the caller to quickly tell their reason for calling.*** After an exchange of pleasantries, the decision to discuss the subject matter rests on the shoulders of the person who initiated the call. If they are in an idle frame of mind, they may take more time than necessary before discussing the purpose of the call. You can control this by diplomatically asking, "What can I do for you today?" or "What is it you are in need of today?".

If someone calls they must want something so don't hesitate to ask them! Asking someone the reason for their call is not rude if you consider that you have something they want and you are willing to give it to them. An abrupt "What do you want?" is too harsh. However, the earlier you ask the reason for the call, the quicker you will be able to end the conversation. After you have answered their question ask, "Is that all I can do for you today?" If the answer is yes, then you have the right to end the conversation and get back to your priorities. If

*It is a prudent use of your time and your resources to make sure the telephone is an instrument of communication instead of an instrument of convenience.*

the answer is no, proceed until you get a yes answer.

The third technique is to **maximize phone conversations when YOU call someone.** There are two time management techniques you can use to improve your telephone protocol. The first is to spend a few seconds and *jot down an outline of why you are calling.* Secondly, *identify the information you want prior to dialing the number.* When they answer you will be able to state, "I need your help on....".

Getting to the point of the conversation and having an outline of the questions you intend to ask, you will not spend a great deal of time searching for what to ask. This is also very respectful of the other person's time. This action on your part reinforces the idea that you like to have your time respected as well.

The telephone company has several clever advertisements stating that the best business calls are personal calls. That's true sometimes. At other times the best business calls are short and to the point. You have heard the phrase, "Time is money." By the same token, telephone calls cost money because they take time. It is prudent use of your time and resources to make the telephone an instrument of communication instead of an instrument of convenience.

The last technique regarding time management is to **control your "turf."** It seems to make some individuals feel powerful to bring people to your office for a meeting. However, when you bring people into your territory it can be difficult to get them to leave. One skill you can develop is to take your communications to the other person's turf. By doing this you have the ability to terminate the conversation once your objectives have been accomplished. It is easier to go to someone else's office, commit yourself to the conversation, and then excuse yourself by saying, "I need to get back to my office to finish something." If you bring the person to your office, asking them to leave can cause them to feel they are being excused from a papal audience.

> *By communicating in another person's office you have the ability to terminate the conversation once your objectives have been accomplished.*

## Saving A Little Means A Lot

These are some basic techniques for improving your use of time. None of them will add hours to the day. However, used collectively they may help you recapture an extra hour or two a day.

If I asked you how much you could accomplish if you were sent into the mountains to concentrate on one project for two weeks, you would probably say "Lots!" If you are able to use thirty minutes a day more effectively, you will gain two and one-half additional work hours per week. This multiplied by fifty weeks is over one hundred extra hours per year. That is more than two weeks of extra time you have found per year to accomplish your work goals. There is no telling what you can accomplish with one hundred hours a year when you set your mind to it!

## DO ONE MINUTE MANAGERS HAVE ENOUGH TIME?

The comedian, Robin Williams, once had a line that went, "Reality — what a concept!" The line got a lot of laughs because for most people reality and their desires are often different. Williams' line capitalized on the humor we find in frustration.

As fire officers we must accept the reality that there are more demands on our time than there are minutes in the day. Time management — what a concept! In the last couple of years author Ken Blanchard introduceded a concept called the *One Minute Manager.* There have been subsequent books called *One Minute Parent* and *One Minute Employee.* How about writing a book entitled *One Minute Fire Officer?*

I am not sure being a one minute fire officer is possible or desirable. If you have read Blanchard's book you are aware he suggests that praise and feedback should be given to individuals in small doses at the time the behavior is observed. That concept has merit, but is sixty seconds sufficient time to capture the essence of the situation or to give an individual the feeling of praise? In actuality, I don't believe Blanchard meant that all praise and feedback should be crunched into a sixty second time frame.

It doesn't take much of a mathematician to realize that in a given day we have 480 opportunities to be a one minute manager. Each work week has 2,400 opportunities to be a one minute manager, and in a year there are over 100,000 opportunities. Mind boggling, isn't it?

In the fire service we can accept the concept of taking slices of time to accomplish specific objectives, but the slices of time need to be more in keeping with the needs of the fire service. The fire service is extremely time-oriented. How do we measure the progress of a fire? By time, of course. How do we measure the effectiveness of our response patterns? By response time. How do we measure our training programs and the amount of time devoted to our various activities, such as public education and fire prevention? By documenting hours.

I am reminded that one of the most critical decision making processes fire officers engage in is called "size-up." We have all had the experience of leaving the fire station knowing we have a response time of approximately four minutes to arrive at the scene of an emergency. Etched against the sky in daylight is a billowing cloud of black smoke. At night the rosy glow of loom-up tells us that we better have our act together when we arrive on the scene. Four minutes. In those four minutes, we try to consolidate an entire career of experience, knowledge, and education to make sound decisions under highly stressful circumstances.

*The concept of taking one minute to give frequent praise and feedback has merit. However, is sixty seconds sufficient time to capture the essence of the situation?*

### Be A Five Minute Fire Officer

Instead of becoming one minute managers, fire officers should become five minute decision makers. We should utilize the same methodology we use in controlling major emergencies. As fire officers, we should take five minutes to consolidate, integrate, and synthesize our knowledge and experience as we approach each management task in our fire departments on a daily basis. We should regularly take five minutes to give feedback and praise to our subordinates.

To be more specific about how a five minute fire officer might function, let's look at a staff meeting. I once was in a staff discussion that was a disaster. As we sat around the table it was obvious we were going to be asked to "participate" in a decision making process. However, the person chairing the meeting displayed body language somewhat like a skeetshooter. You could almost hear the click-clack of a round being loaded into his mental shot gun as he prepared to shoot down all proposed ideas. The problem was laid on the table and solutions were suggested. The ideas fluttered into the sky like skeet clay pigeons. Boom! Boom! Boom! Every idea was abruptly shot down with a statement like, "That won't work," "We've tried that before," or "It's too costly."

A five minute fire officer would not conduct a meeting in this way. Instead, the officer would listen and absorb a new idea, allow it to incubate for a few moments, and see if it generated additional response from other staff members. Instantaneous feedback is far worse than no feedback because it is usually negative. Instantaneous, negative feedback will inhibit the growth of ideas instead of encouraging them.

A five minute fire officer would also sincerely reward people for quality performance. I recall an in-

cident where a chief officer attended a class called "Non-Financial Incentives." Immediately upon return to his office, he called all of his staff together, sat them down, and stated, "You guys listen up because I am going to motivate you." Guess how much motivation actually occurred! When people feel they are given praise only for the purpose of motivating them to put out more work, it is worse than receiving no praise. Instead, the chief might take the time to ask himself, "What caused this person to perform so well?"

A five minute fire officer reinforces **reasons** people engage in positive behavior, rather than the result of the behavior. It would take only sixty seconds to say, "Great job, keep it up!" A few more minutes spent understanding the reasons why the person engaged in the positive behavior might result in additional reinforcement such as, "That new training program you developed is certainly going to be valuable to our recruit firefighters. What do you think we ought to do next? Why hasn't our driver training program been as successful as your new concept?"

A five minute fire officer should take the time to contemplate the reasons for success before acknowledging them. Hastily drawn praise that is too general will be lost in the conversation. On the other hand, well developed, specific acknowledgement of success forms the foundation for continued success.

## Evaluate Before Giving Criticism

One minute criticisms have their disadvantages also. It is best to take the time to understand exactly what is being criticized before criticizing a behavior. I recall an inci-

dent in my career involving a fireground command that my superior severely criticized. I arrived at the scene of a high rise fire approximately ten minutes before the chief.

When I arrived the fire was in the flashover stage and boiling out of a window on the eighth floor. A second alarm was struck. The dispatcher notified the chief, who responded to the scene of the emergency.

By the time the chief arrived the second alarm companies were already on the scene. The fire was totally under control and nothing was showing on the exterior of the building. The street looked like a lot for used fire trucks. As far as the eye could see there were flickering emergency lights. Not a firefighter was in sight, except for one pump operator who was providing water to the standpipe.

Without a moment's hesitation, the chief asked, "What the heck is all this equipment doing committed at the scene?" He was distressed to think we had all these resources not being put to work. Without waiting for an explanation, he gave a one minute dissertation on the effective use of equipment. Unfortunately, he did not take time to evaluate my reasons for making a second alarm. Later the chief got all the information and apologized for being so abrupt. The point is this, one minute feedback requires several minutes of analysis.

The five minute fire officer does not draw fire for effect. The "Ready-Fire-Aim Syndrome" does not result in improved performance. Instead it results in a form of institutionalized paranoia where people are looking over their shoulder to see if the officer is going to criticize an action without a chance to explain his behavior.

Maybe you are a one minute manager or a sixty minute fire officer. In either case, the amount of time we spend directing, counseling, and providing feedback to our subordinates is essential to the overall momentum of our organization. Whether it is done in one minute increments or five minute blocks of time, it is important to apply the principles we have covered.

The only thing that counts is that you have a system that works. With due respect to Ken Blanchard, **we don't need to be clock watchers — we need to be people movers.**

*Well developed, specific acknowledgement of success forms the foundation for continued success.*

# POST-PROGRAM ACTIVITY 1

### THE DAILY TIME LOG

In order to gain control of time, it is necessary to understand what we presently do with our time. To discover this, there is only one adequate tool — the time log. A time log is a complete diary of what we do during each day or shift, recorded along with analytical observations. These observations enable us to evaluate the worth of our activities, how our time is allocated to the most valuable activities and what corrective measures can be taken. Three days should suffice if they are logged in sufficient detail to provide a basic understanding of the allocation of our time and the nature and extent of the interruptions we experience.

Follow these instructions for taking your own daily log:

1. Enter name, date and goals.

2. Record each activity as the day progresses. Record all interruptions, their source and reason. Give as much detail as possible.

3. Set a priority on each action, so that you can check back at the end of the day to see how much time was spent on top priority work.

4. Comment on each action with a view to future improvement. Try to note suggestions for making these improvements.

5. Keep logs for minimum of three days, one week is preferable. Allow time at end to analyze your logs.

6. Use signs and abbreviations.

7. Record all activities.

### Examples:

Emergency response
Meet with _____ in office
Personnel Matters
Reports, correspondence, writing
Coffee, eating, personal

Outgoing & incoming calls; from whom
Inspections
Routine paperwork
Training
Open & read mail

# DAILY TIME LOG

Name:_________________________________ Date:___________________________

Daily Goals:        Deadline:                    Daily Goals:        Deadline:

1.________________ _________        5.________________ _________

2.________________ _________        6.________________ _________

3.________________ _________        7.________________ _________

4.________________ _________        8.________________ _________

**Priority:** 1=Most Important; 2=Less Important; 3=Routine Detail; 4=Least Important

| TIME | ACTIVITY | TIME USE | PRIORITY | COMMENTS |
|---|---|---|---|---|
|  |  |  |  |  |
|  |  |  |  |  |
|  |  |  |  |  |
|  |  |  |  |  |
|  |  |  |  |  |
|  |  |  |  |  |
|  |  |  |  |  |
|  |  |  |  |  |
|  |  |  |  |  |
|  |  |  |  |  |
|  |  |  |  |  |
|  |  |  |  |  |
|  |  |  |  |  |

## THE WEEKLY PLAN

In conjunction with the daily time log which analyzes your every day activities, the weekly plan is a useful tool to plan several days or shifts in advance. The weekly plan is to be done with your company. If you are assigned to a staff function it is recommended that you complete the form with a coworker.

Follow these instruction for developing your weekly plan:

1. Have a company or officer meeting to mutually set goals for the coming week or set of shifts.

2. Prioritize each goal and record the goal statement on your weekly plan.

3. Estimate the time needed to complete each task.

4. List any special activities which will be required to accomplish your objectives. Examples:

- Call to schedule inspections
- Arrange a training site
- Set up a meeting with _______

5. List the objectives to be met each day on your day planner.

6. Review your plan the following week to see if you meet your objectives. If you did not, examine why.

# WEEKLY PLAN

DATE:

| DATE | PRIORITY | TIME REQUIRED | OBJECTIVES | ACTUALLY ACCOMPLISHED |
|------|----------|---------------|------------|------------------------|
|      |          |               |            |                        |
|      |          |               |            |                        |
|      |          |               |            |                        |
|      |          |               |            |                        |
|      |          |               |            |                        |
|      |          |               |            |                        |
|      |          |               |            |                        |
|      |          |               |            |                        |
|      |          |               |            |                        |
|      |          |               |            |                        |
|      |          |               |            |                        |
|      |          |               |            |                        |
|      |          |               |            |                        |
|      |          |               |            |                        |

# POST-PROGRAM ACTIVITY 2

**SCENARIOS - MODULE VI:**

**Scenario 1**

**Setting:** Office

**Company Officer:** Good morning, guys. Here we are again!

**Firefighter One:** What's on the agenda today, boss?

**Firefighter Two:** The usual — hero stuff!

**Company Officer:** You got it! Actually, there are only a few things scheduled. We have a class at 1000, then an inspection appointment at 1300. Most of the morning will be taken up by the class, but the afternoon is open. Do you guys want to do anything in particular?

**Firefighter One:** I've been trying to get in touch with that guy in P.D. to set up some training.

**Firefighter Two:** I'm supposed to be working on that equipment proposal for staff.

**Company Officer:** I'm a little behind on filling out your evaluations. Sounds like we all have some individual projects we could be working on. Let's spend the later part of the afternoon on those items, okay?

**Firefighter One:** Okay, fine.

**Firefighter Two:** Another day in the life.... Let's have some coffee!

---

Evaluate this scenario and describe the positive and/or negative content. If you were the company officer, would you have done anything differently?

_______________________________________________

_______________________________________________

_______________________________________________

## Scenario 2

**Setting:** Day Room

**Firefighter One:** Here we are again.

**Firefighter Two:** Yep, always ready to handle a crisis. I wonder what HE has in store for us today.

**Firefighter One:** Who knows. He'll probably let us know when it's time to get on the engine to go.

**Firefighter Two:** Yeah! How's that R&D project going that you've been working on?

**Firefighter One:** Not very well. I don't seem to have enough time. Seems like every time I get started, it's time for us to go somewhere. Like last shift, I had just spread out my drawing when that tour showed up! It would sure be easier if we knew what was planned for the day.

**Company Officer** (walking into the office): Hi, guys. What's up?

**Firefighter Two:** That's what we were wondering!

**Company Officer:** Oh, you mean today? Not much, just the usual I guess. You know — training, inspections — all of the regular stuff.

**Firefighter One** (glancing at Firefighter Two with a knowing look): Yeah, we know.

———————————

Evaluate this scenario and describe the positive and/or negative content. If you were the company officer, would you have done anything differently?

———————————————————————

———————————————————————

———————————————————————

## Scenario 3

**Setting:** Office

**Company Officer:** I just received next month's calendar. It looks pretty open. Only two classes are scheduled.

**Firefighter One:** Great — lots of time to get our own things done.

**Company Officer:** Let's make a list of what we want to accomplish, then prioritize the list and put them on the calendar. That way we'll know what kind of a time schedule we have to keep in order to get everything done by the end of the month. Bob, what to you want to accomplish this month?

**Firefighter One:** I'd like to spend some time getting used to the new extrication equipment we received. I'm also in the middle of wage negotiations, so I could use some time on the phone with other departments.

**Company Officer** (making notes): Okay Jack, how about you? Do you have any ideas?

**Firefighter Two:** I've been wanting to review my hydraulics. You know, all of the rules of thumb, fire flow problems, friction loss — that sort of stuff.

**Company Officer** (continuing to take notes): Okay, let's see — training, hydraulics, and extrication. Paperwork — Bob on the phone. Bob do you think two afternoons will be enough?

**Firefighter One:** Yeah, should be.

**Company Officer:** Okay, I need to meet inspection minimums. I should be able to get that done in two half days. Jack, would you be willing to prepare a hydraulics class for us? I'm sure we could all use a refresher. How about the first shift of the last set? No big deal, just cover what you want to review.

**Firefighter Two:** Sure. That would give me enough time to prepare.

**Company Officer:** Great! Between departmental scheduled classes, the inspections, extrication, and hydraulics classes of our own, this month ought to fly by. Are we leaving anything out?

(Both Firefighters shrug their shoulders.)

**Company Officer:** Let's call it a plan, then!

Evaluate this scenario and describe the positive and/or negative content. If you were the company officer, would you have done anything differently?

_________________________________________________________________

_________________________________________________________________

_________________________________________________________________

# REFERENCES — MODULE VI

Bliss, Edwin C. *Getting Things Done.* Bantam Press, 1976.

Committee for Economic Development, Research and Policy Committee. *Improving Productivity in State and Local Government.* 1976.

Lakein, Alan. *How to Get Control of Your Time and Your Life.* The New American Library, Inc., 1973.

Mackenzie, R. Alec. *The Time Trap.* McGraw-Hill, 1975.

Oncken, William Jr. *Managing Management Time.* Prentice Hall, 1984.

Ross, Joel E. *Managing Productivity.* Reston Publishing Co., 1977.

# MODULE VII

## COACHING FOR TOP PERFORMANCE:

## EVALUATIONS

**"When you're through improving yourself, you're through."**
**— Rex Golobic**

# MODULE VII NOTES

# MODULE VII

# COACHING FOR TOP PERFORMANCE: EVALUATIONS

## OBJECTIVES

To provide the viewer and participant with:

- a basic understanding of career development concepts
- a basic understanding of counseling and interviewing techniques
- a basic understanding of the concept of discipline.

## OUTLINE OF KEY POINTS

**Career Development Concepts** — Each individual needs a career plan showing how they will get from where they are now to their career goal (see Figure 7-1). Career potential can be developed using one of two methods: 1) be dependent upon individual efforts or 2) be dependent upon organizational support and guidance. These methods are incorporated in the following career development approaches:

- **Self-Help Approach** — Low organizational influence and low dependence upon others. The self-help method of career development requires a strong self-image. The disadvantage is the individual relies only on what they know. If everyone used this method teamwork would be hampered.
- **Network Approach** — Low organizational influence, but high dependence on others. Firefighters share information. Organizational goals are a low priority, concern is with what peers are doing. The advantage is that working for a common goal builds good feelings between peers. Disadvantage is if this method is in full effect, it tends to ignore individual goals and focus on group goals.
- **Mentor Approach** — Low dependence on others and high organizational influence. Someone within the organization picks an individual to teach what they know. This method allows the organization to hand pick who will move up through the ranks. This method is used when the organization is structured and task oriented. The advantage is it focuses on skill development of selected individuals which causes the transition to the officer role to be less threatening. The disadvantage is that it can be unfair. What criteria will be used to select who will to be chosen for the mentor program?

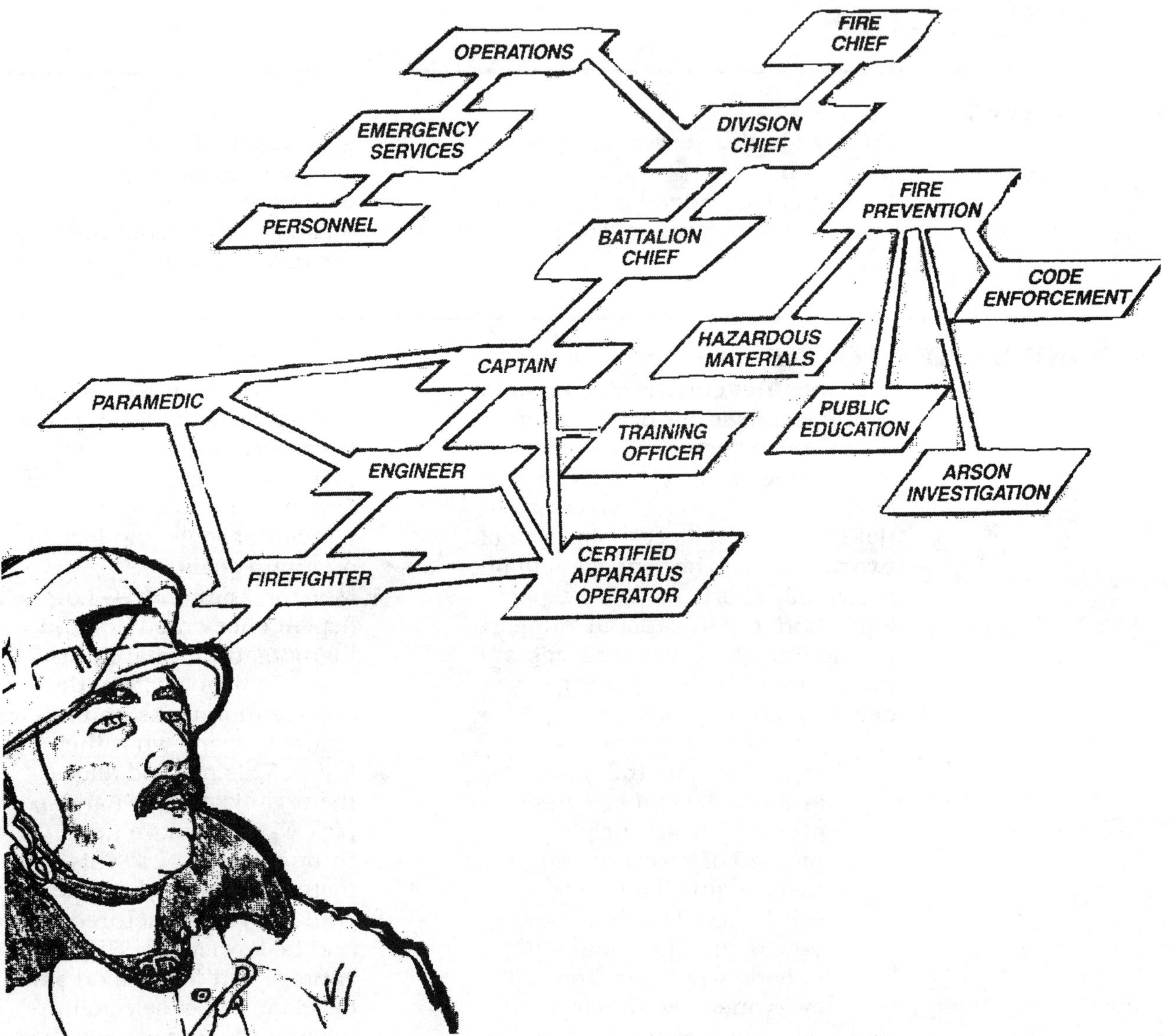


**Figure 7-1   Career Planning**

- **Engineering Approach —** High organizational influence and high dependence on others. This approach synthesizes the best of the self-help and organizational methods of career development. The organization makes employees aware of what they need to do to develop their career and the individual makes choices accordingly.

**Career Development Guide —** The Career Development Guide is a document that provides parameters for successful advancement within the organization. It clarifies both the requirements of the job and the methods of achieving necessary skills and abilities. A Career Development Guide brings into focus what the organization expects and at the same time allows the employee to plan his/her own career.

**Performance Analysis —** A procedure for identifying the cause, importance, and potential solutions for a personnel problem. It is an analysis useful for problems that are the result of inappropriate or inadequate human performance. Performance analysis isn't a criticism of the person, it is simply looking over how the person is performing. During the analysis standards should be set up and expectations of the employee should be made clear. Performance analysis may include the following:

- **Performance Evaluations** — The first step to performance appraisal is to define the job and the criteria by which performance will be judged.

- **Communicate Constantly** — Performance evaluations are based on everyday activities. Feedback should be given continuously regarding performance, not just once a year.

- **Collect Information** — It is important that we collect specific incidents to support performance assessment. These should be documented as they occur rather than being recalled months later.

- **Approaching Evaluations** — The performance evaluation is an important event and should be conducted in a professional manner. The atmosphere should be kept positive, even if the evaluation contains areas for needed improvement.

- **Positive Discipline** — Discipline is derived from disciple. Behavior, both positive and negative, can be learned and reinforced. Positive discipline is almost entirely derived through training and counseling that reinforce appropriate behaviors.

- **Handling Problem Employees** — In a large percentage of cases, problem employees are caused by factors other than personality. Identification of the causes of behavior is extremely difficult. Supervisors should prepare individuals for feedback and counseling before a behavior becomes a problem.

# PERFORMANCE PLANNING AND CAREER DEVELOPMENT

## Benefits of Performance Planning —

- Employees will have a higher level of committment to their work.
- Fewer false steps are taken if a clear course is set.
- Provides needed support and resources to get the job done.

## How To Plan —

- Always write your plans out, don't just think about them.
- Begin with overall organization or department goals.
- Get your employees involved in the planning of their own work.
- Set goals and plans based on overall organization objectives.
- Add work described in the employee's job description.
- Modify plan based on the employee's strengths and limitations.
- Modify plan based on limits imposed by the environment in which the employee will work.
- Make detailed action plan, include actions the manager will take to support the employee's work.

## Brainstorming May Help Decide What Work To Include In The Plan —

- First list everything you can think of regardless of whether it is practical.
- After you have listed all your ideas, prune the list to those that are most productive.

## Benefits of Employee Involvement in Planning —

- Their knowledge and perspective will make the plan more realistic.
- They are more likely to commit to the plan with enthusiasm.
- They are less likely to resent being assigned certain tasks if they know why they are doing them and how they fit into the larger picture.

## Criteria for a Good Career Development Plan —

- Be specific.
- Goals should be measurable.
- Have deadlines.
- Be realistic.
- Work should be challenging.
- Work plans should include objectives, as well as action

plans and statements about the limits of an employee's authority. Make your own action plan stating how you can support the employee and follow-up dates.

## Include Personal Growth Objectives in the Plan —

These are the employees' responsibility. They can include:

- Set goals for performance improvement.
- Career related goals.

## Set Performance Improvement Goals That —

- Relate to the person's work.
- Are built into the regular work plan, if possible.

## Career Goals Should Be —

- Flexible.
- Both long and short range.
- Set standards of performance high and apply them fairly and equally to all employees.

## Relate Overall Standards of Performance to Salary —

The following is a possible rating system which can be used to measure performance:

- Exceeds most of the time.
- Exceeds some of the time.
- Meets all standards.
- Meets some standards.
- Falls below standards.
- Your own enthusiasm for the plan sets the tone for the employee's response.

## Suggested Checklist for Counseling Employees —

- Stop talking and start listening.
- Carry on only one conversation at a time.
- Empathize with the person speaking.
- Don't be afraid to ask questions.
- Don't interrupt the other person.
- Show interest in their needs.
- Concentrate on what is being said, not how it is being said.
- Don't jump to conclusions.
- Control your anger.
- React to ideas, not to the speaker.
- Listen for what is not said.
- Share the responsibility for communication.
- Evaluate facts and evidence on their merit.

# INVESTING IN YOURSELF

Have you had enough of cliches? We have been exhorted to "search for excellence," become "one minute managers," we have gone through "future shock," and we have ridden the "third wave."

It seems like every time we go away to a conference, or worse yet, when our superiors come back from a conference, we are given a whole new set of buzz words and cliches regarding behavior and future opportunities. In the field of public administration it seems that there are a lot of people who lie awake nights to come up with new theories about how we should behave. Quite frequently some of these cliches are turned around and used as means of evaluating productivity and performance.

For example, Management by Objectives (MBO) evolved from a relatively simple concept into a massive paperwork monster in many organizations. Another example is zero based budgeting. For many organizations zero based budgeting meant, "what the city giveth — the city shall taketh away." We need to develop a defense against this barrage of acronyms. How can we best turn this information into an asset for the fire officer?

## We Need Advanced Warning

If we have a hawk's eye view of what is coming through the communications network of federal government and private enterprise, we may have advance warning of what will be landing on our desks someday. It is important for us to recognize that the average fire officer believes most of his information will come to him through professional organizations. In other

The simplest technique that can be used to control our destiny is information. Information is power. The more you know about "the way things are," the more likely you are to be able to do things your own way. My suggestion to deal with this is so simplistic that it is probably overlooked by most of us on a daily basis. The solution — **invest in yourself!**

If you take a portion of your budget and invest it into an information network, it will allow you to know things before they come from the top of the bureaucracy down to you. If a portion of your discretionary funds is used to obtain a wide variety of subscriptions and memberships, you will be handed valuable ammunition. I am talking about "information ammunition."

Many city managers and politicians are trying to make government work more like a business. Most of the terms and buzz words that have come down through government bureaucracy have first had their heyday in the field of private enterprise. They have been experimented with first in private enterprise, then by the federal government before they ever arrive on the desks of local government.

words, they rely a great deal on professional publications such as *Fire Chief, International Association of Fire Chiefs, Fire Command, Fire Engineering, Fire Journal*, the *International Society of Fire Service Instructors, Rekindle*, the *American Fire Journal, Fire House*, etc.

Further, we have many state and regional fire magazines whose columns repeat the latest "dance

> *If a certain portion of your discretionary funds is used to obtain a variety subscriptions and memberships, you will be handed "information ammunition."*

176

craze." Our professional publications do not provide us with much advance warning of changes we may expect. It would be ideal if each fire department had access to all of the publications. All of the magazines listed above can be subscribed to for a total investment of less than fifty cents per working day.

## Look To The "Outside" For Help

Fire officers need to expand their scope of information sources a little. The early warning system for the Strategic Air Command is a radar system used to detect our enemies when they are still out on the front lines. They are not next door. The Strategic Air Command calls it the DEW lines. This stands for Distant Early Warning. How about setting up a Distant Early Warning device for yourself?

One tool that is available is to engage in a "frequent buyer" program in the management textbook area. There is actually a group called the "Management Books Institute," which is a type of "Book of the Month Club" for people in management. You can receive publications as they come on the market. Each month you may purchase the book that is the monthly selection, choose another book, or make no purchase. There are no obligations to buy any particular number of books as a member of the Management Books Institute. The book price ranges from $12.00 to $15.00. Some of the books are basic, others are on the best seller list. For example, one of the recent offerings from this group was entitled, *What They Failed to Teach You at Harvard Business School.*

To join the Management Book Club you should drop them a line on letterhead or with your business card to:

Management Books Institute
Route 59 at Brook Hill Drive
West Nyack, New Jersey 10994

If you were a medical doctor or an attorney you would spend great sums on journals and subscriptions to keep your library current. Book expenses for professional purposes are tax deductible. You get a double asset — information and a tax deduction.

Another early warning mechanism is a magazine entitled, *Success. Success Magazine* is edited by W. Clement Stone. Mr. Stone is one of the "gurus" of management methodology in the business arena. This monthly magazine focuses on success stories in business and industry. It provides articles on trends and patterns in management methodology.

*Success* carries a variety of advertising that can be used as a source to keep abreast of changes in corporate and business philosophy and direction.

A subscription to *Success Magazine* is approximately $36.00 yearly. This is about the cost of one coffee break a week for an entire year. As mentioned previously, it is a dual asset which provides early warning information and a tax deduction. You can subscribe to *Success Magazine* by writing to:

Success Magazine
P. O. Box 3038
Harland, Iowa 51537

Another excellent resource to keep you aware of what is coming down the pike is a series of publications put out by the International City Management Association (ICMA) entitled "Management Information Services Bulletins" (MIS).

The MIS series focuses on community practices that are on the leading edge of change.

There is one scene in the movie "Patton" in which General Patton was engaged in tank warfare in the African desert. He was reversing the trend of the German Panzer successes by using some of their own tactics against them. Patton's line was, "Rommel, I read your book!" I am not suggesting that we are engaged in combat with city management, however, it is very important that we are on the same wave length they are. Periodic review of the list of MIS bulletins will allow you to access and read the ones that are pertinent. This enables to you stay in a proactive mode.

For example, the entire concept of "quality circles" was introduced in Japanese and some European literature. It also appeared in the ICMA MIS series. The business concept of "strategic planning" has been another topic addressed in the ICMA bulletins. Almost all current and forthcoming management philosophies have been addressed by the ICMA in their trade journals.

It is not necessary to be a member of the ICMA to access this information. You may obtain a list of MIS bulletins by dropping a card to:

Order Department
International City Management Association
1120 "G" Street, Suite 300
Washington, D.C. 20005

## Information Doesn't Need To Be Costly

The last technique in being aware of new information is to send for free information that is listed in professional publications. There is frequently a section in each publication that deals with book reviews or describes current literature. One way to keep your in-basket full of new information is to read these sections carefully and collect as many free publications as you possibly can that are generated by governmental and/or business agencies. I have a standard letter in the word processor of my computer which refers to an issue of a professional publication, asks if I can receive a copy of the article or publication mentioned in that issue, and thanks the organization for making this available. This letter can be sent out with very little effort.

When I was a young training officer I was involved in a discussion about the cost of training. One of my fellow training officers told me, "If you think training is expensive, try NOT training someone and see how expensive THAT gets!" It is the same way with investing in your future. If you think magazine subscriptions or buying management books is expensive, try not doing it and see how long you survive in the administrative and management jungle. According to my records it costs me less than $500.00 a year to access the information I have referred to. That is a little more than one dollar a day. You probably leave more than that in tips during coffee breaks and lunch.

***Don't search for excellence — create it!***

*If you think magazine subscriptions or buying management books is expensive, try not doing it and see how long you survive in the administrative and management jungle.*

---

# FEAR OF FAILURE

In our society, we emphasize winning. Television tells us there is "agony in defeat." Success is associated with the good things in life.

Let's hear it for failure. That's right — FAILURE! Don't get me wrong, I'm not suggesting that failure is a good idea. If you're a coach for a major league baseball team and you consistently fail to win, you won't stay coach very long.

It is important to realize that it is not a disgrace to try something and have it not work. The essence of experimentation is that research and development first produce failures. The failures cause each success to be ecstasy for the experimenter.

Trying and failing is a growth process. How many different experiments did it take for Alexander Graham Bell to invent the telephone? How many substances did Thomas Alva Edison experiment with before he created the incandescent light bulb? How many people crashed kites, gliders, and other imitations of a bird's anatomy before the Wright brothers were successful in getting an aircraft to fly?

Winning is not always important. If you win at things that don't mean much, then the rewards are meager. The greater the risk of failure, the greater the reward.

What does this mean in our everyday world? If we understand failure we will see that our perception of failure deeply effects how we perform as fire officers. If we are interested in seeing our subordinates grow in mental and professional stature, then we must be willing to stand aside and allow them to experience failures of their own creation.

## Encourage Learning Through Experimentation

Let me explain this further. Most fire officers have been involved in the process of helping to build an organization from infancy to maturity. While going through the process many experiences tell you that certain management techniques, leadership styles, and/or programs are not effective.

When a junior officer or firefighter wants to try an idea that experience has shown will probably not work, our first reaction is to discourage it. The act of discouraging the experimentation process is a latent desire on our part to keep our subordinates from needing to walk in our same footsteps and trip over the same obstacles we have. This concern is a double-edged sword.

When we discourage the experimentation and growth of our subordinates, we erode a little of the subordinate's loyalty. It is like a father who tells a child he can't do something and then is confronted with the ever present, "Why?". The response is often, "Because I say so." Such a response drives a wedge between us and the people we hope to prepare to take over our position of responsibility and authority someday.

This does not imply that we should not engage in a dialogue to counsel individuals from becoming kamikaze pilots or leading bonzai charges into the sunset. However, we need to be aware that subtle discouragement keeps our personnel from stretching and growing by

> *It is important to realize that it is not a disgrace to try something and have it not work.*

trying new techniques and concepts.

The idea the individual has proposed may not even be new. Almost all ideas that come to the forefront in the fire service go through stages of acceptance, implementation, deterioration, obsoleteness, oblivion, and then renaissance.

If you stay in this business long enough, you will see most of the ideas that have been promoted at one time as the "final solution" to a fire problem pass away, only to return later to be considered by a future generation. Program concepts, management ideas, and even some technological ideas are like the ties we have hanging in our closet. If you hang onto the thin ones long enough, they come back in style.

The attack pumper is an example. Today the attack pumper is being discussed as a possible solution to dwindling manpower, resources, and the spiraling cost of apparatus manufacturers. If you look closely at the basic design of fire equipment in horse-drawn days, and even in the day of early mechanization, the chemical engine was that generation's version of the attack pumper. Fire prevention is not new. Fire prevention was part of Colonial America. Built-in fire protection was as much of a requirement in the days of Myles Standish as it currently is under the concept of master planning.

## Are You Helping Or Hindering?

How can we allow our personnel to make their own mistakes and grow with the program? I am not advocating that we allow people to make errors in judgment or operate at less than peak efficiency. To the contrary. I believe it is absolutely essential to the professional development of the future fire officer that he/she not only understands what is being done, but also have a good reason for doing it.

When we tell someone they shouldn't try something because we tried it and failed, we are implying that the concept was a failure. That may not be true. The concept may have failed because it was an idea whose time had not arrived or is was not implemented in the best way.

As future generations come along with new data, skills, and supporting systems, they may take concepts that were borderline at one time and turn them into the cutting edge of change for the fire service. Our job is to not blunt that cutting edge through our intervention. Rather than discouraging young officers from trying, we should be encouraging them to seek broader horizons. We need to encourage them to create their own experiences by testing their reasoning, rather than accepting our prophecy of failure.

Many times there is a fear of failure in the top man of the organization. When you're at the top, some individuals develop a concern that if subordinates try and fail it will cast a shadow across their office door. History does not support that theory.

A classic example of fear of failure can be found by looking back in time at the Edison Company. Before a new engineer was hired he was given a light bulb, sent into a small laboratory, and told to coat the inside of the bulb. The engineers working at Edison Company knew it was impossible to coat the inside of a light bulb formed in a vacuum.

*We need to encourage our subordinates to create their own experiences by testing their reasoning, rather than accepting our prophecy of failure.*

It was like a game. Weeks would go by and finally the young engineer would stagger back into the group and admit his failure to coat the inside of the bulb. Admitting he was a failure automatically gave him the password he needed to be hired. The story reached outside the company. Engineers applying for a position merely went through the motions of attempting to solve the problem. Many would go behind locked doors and wait for an appropriate amount of time to pass. They would then emerge and admit defeat to join the ranks of the "it can't be doners."

This kept occurring until one day they tested a young man who had never heard the story. He had no fear of failure. They gave him the light bulb amd he went into the laboratory. One week later he came out with an evenly coated light bulb which increased reflectivity and usability. Eureka, it could be done! But, it would have never be done by those who had been told it was impossible. They did not try to experiment, they did not develop a rationale to explain their failures. They merely accepted failure.

By retirement, this young engineer had probably developed some prejudices of his own. Chances are he shared these prejudices with the next generation of engineers which prohibited the formation of some other idea. That's part of the shame of it all. Creativity and the ability to resolve complex problems periodically requires a fresh approach to an age old problem.

## Encouraging Creativity — A Key To Success

In the role of fire officer, it is important for us to support the efforts of our personnel to bring fresh, new approaches to the problems we face. We should do everything in our power to discourage "yes" men, who merely reinforce our own perspectives. We should encourage our future fire officers to think for themselves. It is easy to say this, but hard to see it through within the fire department. Make no mistake, whenever you encourage creative thinking behavior, the organization becomes increasingly difficult to manage. The risks are great, but the benefits are enormous.

Take a look at what happened to the British Army when it fought the rebels in the 1700's. At that time the British Army was one of the most highly disciplined, highly structured, and inflexible organizations in the military world. Wearing red coats and lined up shoulder to shoulder with fixed bayonets, the British Army had managed to conquer almost every army it had faced. That is, until it ran into a group of sharpshooters wearing coonskin hats and hiding behind rocks and trees. Instead of lining up like robots, these sharpshooters fired at individual targets from their hiding places. The American Continental Army systematically defeated an organization with hundreds of years of "military" experience.

We may draw a comparison in the modern fire service. Highly disciplined, highly organized, and rigid organizations are being shot at from all angles. The ranks are thinning. The casualties are high.

"Lean and mean" organizations are not afraid to try something new. They are not afraid to engage in a little guerrilla warfare of their own. These organizations have retained their flexibility and, therefore, they survive in the face of a changing environment.

Winning isn't everything, that's true. However, learning how to win and keep winning is an important skill in the survival of any profession. As fire officers, it should be our mission to make sure we keep on winning by periodically reassessing our old game plan and trying a new plan from time to time.

We can best avoid catastrophic failure of our organization by allowing our subordinates to learn through doing. That strengthens them in preparation for the bigger battles that will be faced.

The next time one of your officers has a new idea that reminds you of a failure you once had, don't start by talking. First, take the time to listen thoroughly to the idea. If the possible risk is not severe and if the individual has the potential for growing into the kind of leader the future needs, invest a little time and allow him to experiment. Properly invested, the dividends will be returned many times over in the future.

# POST-PROGRAM ACTIVITY 1

## SCENARIOS - MODULE VII

### Scenario 1

You are the Fire Marshal of the department. After extensive negotiation with a contractor over a plan check problem, you finally resolve all of the problems. The contractor did not want to sprinkler the building, yet it was oversize. A compromise was achieved by installing division walls. This was accomplished by your suggestion only after weeks of disagreement.

A week later when you arrive in the office you discover an envelope on your desk. Upon opening it. you discover two tickets to a professional baseball game and a note stating, "Thanks for all your help." Do you:

a. take the tickets to your superior's office and report them to him,

b. call the contractor up and ask him to come over and take the tickets back,

c. throw the tickets away and don't mention it to either your superior or the contractor,

d. Use the tickets for the game?

Comments:_________________________________________________

_________________________________________________

_________________________________________________

_________________________________________________

## Scenario 2

You are coming home from a meeting late one night and you are in a fire service vehicle. As you are waiting at a light, a vehicle comes along side of you. You observe an off-duty Captain from your department driving the car. He has not noticed you because he has his attentions on a woman in the seat next to him. They are embracing in a very amorous fashion. As the light changes, the Captain straightens up and you recognize the woman as the wife of another on-duty Captain. Both appear to have been drinking. Do you:

a. honk your horn, get their attention so they see you, and then drive off,

b. honk your horn, get their attention, then motion the vehicle to the curb and warn the Captain right there of the consequences of his actions,

c. allow them to drive off unaware of the fact that you observed them, then talk to the Captain later when he is on duty,

d. allow them to drive off unaware of the fact that you observed them, then you forget the entire matter?

Comments:__________________________________________________

_______________________________________________________

_______________________________________________________

_______________________________________________________

## Scenario 3

You are the department Training Officer. A man walks into your office and identifies himself as an FBI agent. He has a warrant for the arrest of one of your new recruits. The warrant is for narcotics trafficking with a high school student. The recruits are due to graduate in three days. The recruit in question is number one in the class in performance. The Chief is out of town and unavailable. The agent wants the man brought to the office. Currently the recruit is at the drill tower. Do you:

a. immediately contact the local police department and ask them to accompany or meet you at the drill tower,

b. go to the tower yourself, pick up the recruit in question, and return him to the FBI agent without advising him of what is happening,

c. go to the tower yourself, remove the recruit from class, advise him that he is suspended from duty pending an investigation, then take him to the agent at headquarters,

d. ask the agent to accompany you to the tower, remove the recruit from class, and turn him over to the agent,

e. other (please specify)?

Comments:________________________________________________

________________________________________________

________________________________________________

________________________________________________

# POST-PROGRAM ACTIVITY 2

### Pre-Performance Review

The following performance review/goal setting activity is designed to provide communication between the fire officer and firefighter. Although you may already have an existing evaluation system, this review/goal setting activity can provide a non-threatening avenue to review not only your firefighters, but officers as well. We would encourage that you approach this exercise as a learning experience and an extension of the team building concepts outlines in Module V. The purpose is to have the firefighters and officers identify areas of current or potential conflict by comparing Review forms they have completed on themselves and one another. One Firefighter Review form should be completed by the firefighter and another by his/her officer. The Fire Officer's Review form should be completed by the officer and one should be completed by the firefighters under his command. The Review forms should be evaluated during the Performance Review meeting. Differences should be identified and discussed.

Permission is hereby granted by Emergency Resource to reproduce the following Performance Review forms for in-house performance review use only. Duplication for any other use, including resale, is a violation of the copyright law.

# PERFORMANCE REVIEW AND GOAL SETTING INFORMATION
# FOR THE FIREFIGHTER

*Instructions:*

1. Read all of the information on this sheet.

2. Each firefighter and his/her officer should complete a separate Firefighter Review form.

3. Set a time to hold the Review.

4. Conduct the Performance Review/Goal Setting meeting.

**The Performance Review session will follow four (4) steps:**

**Step 1 — Discussing skills that are important to the firefighter's job.**
You and your officer will compare which skills you each feel are most important, important, and least important to your job. How is this helpful to you? If, for example, you know that producing a certain quantity of work is more important than delegating tasks, you can then devote more of your efforts to producing this quantity.

**Step 2 — Discussing the job skills you perform well.** Both you and your officer will talk about the skills you do particularly well. By knowing your strengths, it will be easier to maintain top performance in those areas — and to strengthen your performance in areas you feel need developing. In other words, build on strengths.

**Step 3 — Planning improvements.** This is an extremely important step in the performance review process, because it is here that you and your officer will set goals that will make your job better.

Here's how it works:

Ask yourself what your officer can do to make your job situation more pleasant and productive for you. Then list three goals you want your officer to accomplish before the next Review period that will make your job better. For each goal you should:

**State Specifically What You Mean —**

For example, if you want your officer to "have better communications with me," what do you mean? Should your officer give you more feedback, clearer oral instructions, more frequent written instructions, or say hello to you in the morning?

## Explain How This Goal Is To Be Achieved —

What **exactly** would you like your officer to do in order to reach the goal? Again, if your goal is better communication, do you want a regular meeting with your officer? More frequent memos and reports from your officer? Or do you want other kinds of communication?

## State How And When Progress Will Be Measured —

**How** will you and your officer know if both of you are making progress with each goal? **When** will you monitor results? You should establish checkpoints to do this. You can then let your officer know if you are receiving the kind of communication you want and need. Finally, put down the date you realistically think this goal can be accomplished.

**Step 4 — Agreeing on improvement goals.** During the Review meeting, you and your officer will discuss the goals each of you has established for the other. Although each of you will set three goals, you may each want to tackle only two of these goals for this next review period.

After agreeing on goals, you and your officer will write these goals on the Review form. This form should be signed by both you and your officer. Keep one copy in your files, another for the officer's files, and send a third copy to the Employee Development Department or your administrative offices. Then, at your next Review meeting, you can discuss how well both did in accomplishing your goals.

Employee's Name: _________________________ Date: ___________

Reviewer's Name: _________________________

## FIREFIGHTER REVIEW

### I. Which Skills Are Most Important To Your Job?

*Instructions:* Below are 20 performance skills. Place a:

**3** in front of the skills you think are most important to this job.
**2** in front of the skills that are important.
**1** in front of those that are least important to this job.

*Mark only five (5) skills with a 3.* In other words, no more than five skills should be identified as "most important."

_____ a. Quality of work (performance)

_____ b. Use operating policies prescribed for the job

_____ c. Meet city goals, objectives, and commitments

_____ d. Quantity of work (productivity)

_____ e. Uses time efficiently/effectively

_____ f. Follows through on tasks and activities

_____ g. Apply technical knowledge

_____ h. Generate new ideas and suggestions (initiative)

_____ i. Meet deadlines

_____ j. Effective communication with others

_____ k. Makes timely decisions (judgment)

_____ l. Sees a job to its completion

_____ m. Solves job-related problems

_____ n. Maintain effective work relations with co-workers (teamwork)

_____ o.  Provide quality service (public contact skills)

_____ p.  Proper operation and care of equipment

_____ q.  Safety (work area, equipment)

_____ r.  Attendance (punctuality, availability)

_____ s.  Organize, plan, and schedule work well

_____ t.  _________________________________________

_____ u.  _________________________________________

## II.  Which Skills Does The Firefighter Perform Well?

_____ a.  Quality of work (performance)

_____ b.  Use operating policies prescribed for the job

_____ c.  Meet city goals, objectives, and commitments

_____ d.  Quantity of work (productivity)

_____ e.  Uses time efficiently/effectively

_____ f.  Follows through on tasks and activities

_____ g.  Apply technical knowledge

_____ h.  Generate new ideas and suggestions (initiative)

_____ i.  Meet deadlines

_____ j.  Effective communication with others

_____ k.  Make timely decisions (judgment)

_____ l.  Sees a job to its completion

_____ m.  Solves job-related problems

_____ n.  Maintains effective work relations with co-workers (teamwork)

_____ o.  Provide quality service (public contact skills)

_____ p.  Proper operation and care of equipment

_____ q.  Safety (work area, equipment)

_____ r.  Attendance; punctuality; availability

_____ s.  Organize, plan, and schedule work well

_____ t.  ___________________________________________

_____ u.  ___________________________________________

## III.  Setting Improvement Goals

How can your officer assist you during this next year? List three goals your officer could do that would help you be a more productive firefighter. Remember that an effective goal:

- is realistic,
- is specific about what is desired and how it will be accomplished,
- has checkpoints for monitoring results,
- has a target date for completion.

With these thoughts in mind, state three (3) goals you need from your officer in order for you to be more productive.

**GOAL #1**___________________________________________________

___________________________________________________________

How is this goal to be achieved? ________________________________

___________________________________________________________

How and when will progress be measured? __________________________

___________________________________________________________

Expected date for accomplishing this goal: _________________________

**GOAL #2**________________________________________________

______________________________________________________

How is this goal to be achieved? ______________________________

______________________________________________________

How and when will progress be measured? __________________________

______________________________________________________

Expected date for accomplishing this goal:______________________________


**GOAL #3**________________________________________________

______________________________________________________

How is this goal to be achieved?______________________________

______________________________________________________

How and when will progress be measured?__________________________

______________________________________________________

Expected date for accomplishing this goal:______________________________

# PERFORMANCE REVIEW AND GOAL SETTING INFORMATION
# FOR THE FIRE OFFICER

*Instructions:*

1. Read all of the information on this sheet.

2. Each officer and his/her firefighter's should complete a separate Fire Officer's Review form.

3. Set a time to hold the Review.

4. Conduct the Performance Review/Goal Setting meeting.

**The Performance Review session will follow four (4) steps:**

**Step 1 — Discussing skills that are important to the fire officer's job.**
You and your firefighter will talk about the skills that are most important to your job. How is this helpful? If, for example, you agree that producing a certain quantity of work is more important than delegating tasks, the officer can then devote more effort to producing this quantity.

**Step 2 — Discussing the job skills the officer performs well.** Both you and your firefighter will talk about the skills you do particularly well. When the officer knows his/her strengths, it is easier to maintain top performance in those areas. In other words, build on strengths.

**Step 3 — Planning improvements.** This is an extremely important step in the performance review process, because it is here that you and your firefighter will set goals with one another that will make each of your jobs better.

Here's how it works:

Ask yourself what this firefighter can do to make your job more productive. Then list three goals you want this firefighter to accomplish before the next Review period that will make your job better. For each goal you should:

**State Specifically What You Mean —**

For example, if you want this firefighter to "make better use of work time," what exactly do you mean? Do you want the firefighter to come to work earlier? Do less socializing on the job? Spend more time on important tasks? Delegate more work to others?

**Explain How This Goal Is To Be Achieved —**

What **exactly** would you like your firefighter to do in order to reach the goal? Again, if your goal is better use of work time, do you want this person to learn how to organize tasks more efficiently? In addition, does the firefighter have adequate resources to meet this goal?

**State How And When Progress Will Be Measured —**

**How** will you and your firefighter know if both of you are making progress with each goal? **When** will you monitor results? You should establish checkpoints to do this. At this time, you and the firefighter can discuss whether or not you feel that person is making better use of work time. Finally, write down the date you think this goal can realistically be completed.

**Step 4 — Agreeing on improvement goals.** During the Review meeting, you and your firefighter will discuss the goals each of you has established for the other. Although each of you will have set three goals, you may each want to tackle only two of these goals for this Review period.

After agreeing on goals, you and your firefighter will write these goals on the Performance Review form. This form should be signed by both of you. Keep one copy in your files and send a second copy to the Employee Development Department or your administrative offices. Then, at the next meeting, you can discuss how well both of you did in accomplishing your goals.

194

Employee's Name: _________________________ Date: _____________

Reviewer's Name: _________________________

## FIRE OFFICER'S REVIEW

### I. Which Skills Are Most Important To The Officer's Job?

*Instructions:* Below are 20 performance skills. Place a:

**3** in front of the skills you think are most important to this job.
**2** in front of the skills that are important
**1** in front of those that are least important to this job.

*Mark only five (5) skills with a 3.* In other words, no more than five skills should be identified as "most important."

_____ a.  Quality of work (performance)

_____ b.  Use operating policies prescribed for the job

_____ c.  Meet city goals, objectives, and commitments

_____ d.  Quantity of work (productivity)

_____ e.  Uses time efficiently/effectively

_____ f.  Follows through on tasks and activities

_____ g.  Apply technical knowledge

_____ h.  Generates new ideas and suggestions (initiative)

_____ i.  Meets deadlines

_____ j.  Effective communication with others

_____ k.  Makes timely decisions (judgment)

_____ l.  Sees a job to its completion

_____ m.  Solves job-related problems

_____ n.  Maintains effective work relations with co-workers (teamwork)

______ o.  Provide quality service (public contact skills)

______ p.  Proper operation and care of equipment

______ q.  Safety (work area, equipment)

______ r.  Attendance (punctuality, availability)

______ s.  Organize, plan, and schedule work well

______ t.  _______________________________________

______ u.  _______________________________________

## II. Which Skills Does The Officer Perform Well?

______ a.  Quality of work (performance)

______ b.  Use operating policies prescribed for the job

______ c.  Meet city goals, objectives, and commitments

______ d.  Quantity of work (productivity)

______ e.  Uses time efficiently/effectively

______ f.  Follows through on tasks and activities

______ g.  Apply technical knowledge

______ h.  Generates new ideas and suggestions (initiative)

______ i.  Meets deadlines

______ j.  Effective communication with others

______ k.  Makes timely decisions (judgment)

______ l.  Sees a job to its completion

______ m.  Solves job-related problems

______ n.  Maintains effective work relations with co-workers (teamwork)

______ o.  Provide quality service (public contact skills)

______ p.  Proper operation and care of equipment

_____ q.  Safety (work area, equipment)

_____ r.  Attendance (punctuality, availability)

_____ s.  Organize, plan, and schedule work well

_____ t.  ________________________________________

_____ u.  ________________________________________

## III.  Setting Improvement Goals

How can this firefighter assist you during this next year? List three goals this firefighter could do that would help you be a more productive officer. Remember that an effective goal:

- is realistic,
- is specific about what is desired and how it will be accomplished,
- has checkpoints for monitoring results,
- has a target date for completion.

With these thoughts in mind, state three (3) goals you need from this firefighter in order for you to be more productive.

**GOAL
#1**________________________________________________________

________________________________________________________________

How is this goal to be achieved?________________________________

________________________________________________________________

How and when will progress be measured?__________________________

________________________________________________________________

Expected date for accomplishing this goal:_______________________

**GOAL
#2**______________________________________________

_______________________________________________

How is this goal to be achieved?_______________________

_______________________________________________

How and when will progress be measured?_______________

_______________________________________________

Expected date for accomplishing this goal:_______________


**GOAL #3**__________________________________________

_______________________________________________

How is this goal to be achieved?_______________________

_______________________________________________

How and when will progress be measured?_______________

_______________________________________________

Expected date for accomplishing this goal:_______________

# REFERENCES — MODULE VII


Fournies, Ferdinand F. *Coaching for Improved Work Performance.* Van Nostrand Reinhold Co., 1978.

Mager, Robert F. & Pipe, Peter. *Analyzing Performance Problems or 'You Really Outghta Wanna.'* Fearon Pitman Publishers, Inc., 1970.

**This page intentionally left blank.**

# MODULE VIII

## PROFESSIONALISM:

## A ROADMAP TO ACHIEVEMENT

"There is no such thing as merely surviving or maintaining the
status quo in business. As in the organic world, there is only
growth and decay, and growth is the business of business."
— Isay Stemp, Corporate Growth Strategies

# PRE-PROGRAM ACTIVITY 1

## Alphabet Soup: Organizational and Concept Awareness

*Instructions:* Write out the complete name of the organization or concept that is suggested by the acronym. Each of these relate to the fire profession in some significant way. They are not in alphabetical order.

NBS_______________________________________________________________

NIOSH_____________________________________________________________

NTSB______________________________________________________________

FEMA______________________________________________________________

NFA_______________________________________________________________

DOD_______________________________________________________________

DOT_______________________________________________________________

OSHA______________________________________________________________

NFPA______________________________________________________________

NFPA (SECOND VERSION)______________________________________________

ANSI______________________________________________________________

ICMA______________________________________________________________

UL________________________________________________________________

FM________________________________________________________________

IFSTA_____________________________________________________________

ISFSI_____________________________________________________________

IAFC______________________________________________________________

IAFF______________________________________________________________

IAAI______________________________________________________________

ISO_______________________________________________

EMT_______________________________________________

MCA_______________________________________________

CHEMTREC__________________________________________

ATF_______________________________________________

ICS_______________________________________________

DOJ_______________________________________________

EOC_______________________________________________

SOP_______________________________________________

ZBB_______________________________________________

# PRE-PROGRAM ACTIVITY 2

*Instructions:* The following is a list of professional publications that are offered to the fire service. There are two places on the form for your consideration. The first is for you to check if the publication is available for you to read and be informed. The second is for you to indicate that you read it on a regular basis.

|  | **AVAILABLE** | **READ?** |
|---|---|---|
| Fire Command | _________ | _________ |
| Fire Engineering | _________ | _________ |
| Fire Chief | _________ | _________ |
| Fire Journal | _________ | _________ |
| Fire Technology | _________ | _________ |
| Communication Link | _________ | _________ |
| Rekindle | _________ | _________ |
| American Fire Journal | _________ | _________ |
| Fire House | _________ | _________ |
| Emergency Product News | _________ | _________ |
| National Cities | _________ | _________ |
| Connections | _________ | _________ |
| On-Scene | _________ | _________ |
| State Fire Chiefs Publication | _________ | _________ |
| State Firemans Association | _________ | _________ |

# MODULE VIII NOTES

# MODULE VIII

# PROFESSIONALISM: A ROADMAP TO ACHIEVEMENT

## OBJECTIVES

To provide the viewer and participant with:

- a summary of previously discussed leadership and management concepts as used in the fire service
- a basic understanding of command presence, profes-sionalism, and other skills that are essential to being an effective fire officer.

## OUTLINE OF KEY POINTS

**Attitudes** — Personal percep-tions form attitudes. They can be a springboard to move forward or a trap to keep us where we are. We all have the capacity to use the Pygma-lion or self-fulfilling prophecy to our advantage or disadvantage.

**The Achiever in Our Society** — Even champions have to be trained to achieve. Conditioning is an ele-ment of performance, so it follows that conditioning is a part of our role as fire officers.

**Achieving Credibility** — The badge is not our credibility. Our credibility comes from what we have to contribute to our profession and organization. Credibility comes from:

- Knowledge — both formal and informal, theory and practicality. Knowledge holds a key to power.
- Ability to use knowledge in the real world.
- Ability to communicate with others.
- Ability to understand oth-ers.

**Identification of Your Own Fu-ture** — We start by targeting those skill/abilities that we need to change for more effective relation-ships. This process includes:

- identifying what we would like to see change in our-selves
- identifying what we would like to accomplish in our role
- writing down a "game plan"
- actively searching for new information and experi-ences.

**Developing Your Personal Style** — We don't have to change what we are. We have to change how we use our unique features to get others to achieve. When we cause others to achieve, we also achieve. Style is unique, but each style can be made more effective and credible with study and feed-back.

**Resources and Reinforcement** — We all need a personal resource system. It includes, but is not lim-ited to:

- Educational opportunities
- Training opportunities
- Reading materials
- Role models
- Mentors
- A network
- Personal experiences

## MODELS WE HAVE DISCUSSED

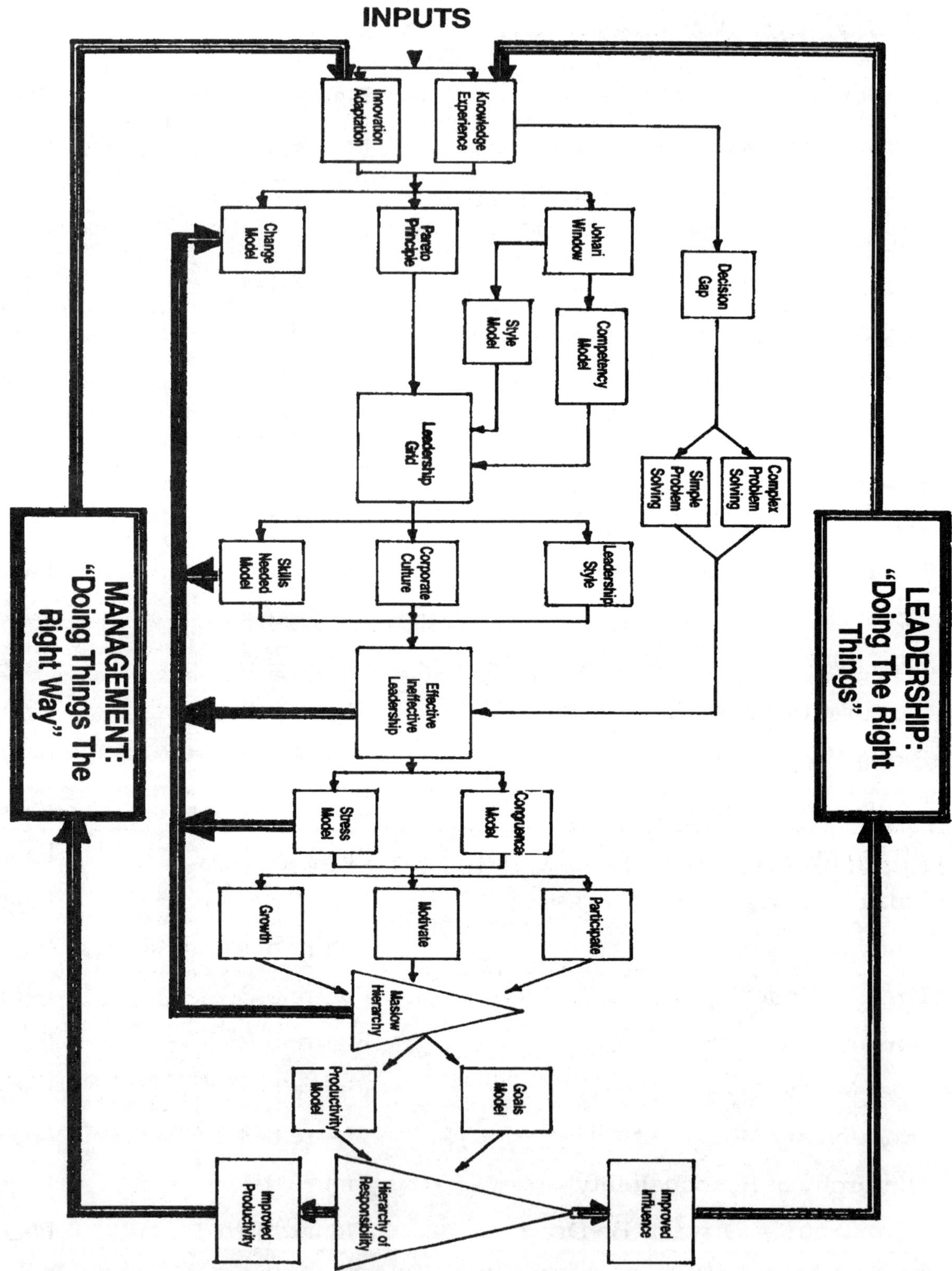


**Figure 8-1    Systems Approach to Professionalism**

## DEFINITION OF A PROFESSIONAL

Webster defines profession as "a calling requiring specialized knowledge and sometimes long and intensive academic preparation." A professional is defined as "a person engaged in one of the learned professions who conforms to the technical or ethical standards of that profession."

It doesn't matter if you are a paid firefighter, volunteer, industrial, or institutional firefighter. Your conduct and ethics must be directed toward protecting lives and property.

A true professional exhibits the qualities and skills acquired through training. The professional uses knowledge, based on experience to carry out his/her responsibilities.

Your level of professionalism is measured:

- by your superiors and
- by people you serve who observe your conduct and skills performed on the scene of an emergency.

Your skills, conduct, and attitude become the yardsticks used to measure you and your entire organizaiton. They establish the reputation that indicates you are a professional.

To acquire a professional reputation as a municipal fireighter, we start with in-service training programs to develop the qualities and skills required of each individual within his or her own department. As a continuous process to improve and expand these skills, we utilize local, regional, state, and national schools and we obtain further information through publications, magazines, and newsletters.

You have an advantage. You have chosen to become a part of one of the greatest professions on earth. It is, therefore, your responsibility to conduct yourself in an appropriate manner and to acquire the skills, educaiton, and certification that sets you apart and provides you with the reputation of a "professional."

## ASSESSING YOUR OWN STRENGTHS AND WEAKNESSES

Self-analysis allows us to:

- recognize those things we are doing well
- take credit for them
- consolidate our strengths
- discover how we can use our strengths to effectively work toward our goals
- learn to use our strengths to build others up.

At the same time, we must know what our weaknesses are. This allows us to recognize our vulnerabilities and turn our deficits into assets. It is important to recognize what we are, but even more important to recognize what we could be. Developing an action plan for the future helps determine what the future will be.

**"The best way to predict the future is to invent it."**

**- Alan Kay, Apple Computer**

## PYGMALION IN THE FIRE SERVICE

A long time ago in a country far from here, there lived a young man named Pygmalion. He was a prince and sculptor. Pygmalion created a statue of a young lady that was so beautiful he fell in love with it himself.

For a long time Pygmalion stared at the statue. He wished the cold, hard marble would come alive. His wish was granted by the gods. His statue not only came to life, but she possessed all of the qualities Pygmalion had hoped for to match her beauty. As in all good fantasies and mythology, the two of them lived a happy life thereafter.

Thousands of years later individuals studying the behavior of human beings remembered the myth of Pygmalion. They wondered to what extent people are affected by the wishes and desires of others.

### Self-Fulfilling Prophecy

Today there are many students of managerial philosophy and organizational development who wonder, "Is Pygmalion alive today?" They believe he is and have even coined a term to express this behavior, *self-fulfilling prophecy.* Self-fulfilling prophecy means if a person thinks something is going to happen, then it probably will occur. The thought becomes reality in spite of circumstances. A self-fulfilling prophecy suggests that an event will occur simply because a person believes it will. Many times that belief has nothing to do with the actual circumstances or conditions surrounding the event. Self-fulfilling prophecies can have either a positive or negative element. Negative self-fulfilling prophecy occurs when people fear something bad will happen, and it does. Positive self-fulfilling prophecy is when an individual hopes something good will happen and their hope becomes reality.

We need to ask, "Is Pygmalion alive in the fire service today?"

There is evidence that he is. At times the fire service is inundated by waves of pessimism. For example, consider the conversations you have had about the future of the fire service the last couple of weeks. How much was centered around the good things that are happening to the fire service? How much optimism have you found lately? How many predictions about our future have a happy ending?

It would appear that criticism of our occupation is more predominant today than optimism. There is a possibility that Pygmalion and the concept of self-fulfilling prophecy are helping to erode our foundation and attack the credibility of our profession.

Most people have a fear of failure. Nobody wants to fail. On the other hand, we spend a lot of time talking about prophecies of failure. That's exactly what a negative self-fulfilling prophecy is.

> *A self-fulfilling prophecy suggests that an event will occur simply because a person believes it will, not because the belief has anything to do with the conditions or event.*

## The "It Won't Work" Syndrome

Sometimes negative self-fulfilling prophecy is called the "it won't work" syndrome. I'm sure you have heard of it, I know I have. I've even seen it take place at the recruit academy level.

A good example can be found in the selection process to admit people into the fire service. It is increasingly difficult for a young person to become a firefighter. Many communities start their testing process with thousands of candidates. These candidates are put through strenuous physical agility tests, mental examinations, background examinations, etc. The result is that the fire service has its pick of the cream of the crop of young, intelligent, talented individuals. They are capable, motivated, and competent.

Yet, many of the entry level training programs bring these same young people into a hostile environment. They are talked down to. The recruits are given the impression that they can only learn a certain amount of information. They are told there are many things they can't know until they have a lot of time on the job. "Experience will be their teacher."

As an experiment, I once took a group of recruit firefighters and taught them a technique that most firefighters don't know until they are well into the officers' ranks. These recruits, not knowing they weren't suppose to be able to perform this technique, became fairly proficient at it in a short time.

This is not to say that firefighters don't have to learn first things first. We have to know the basics before we can move on to bigger and better things. It is necessary to be a good follower before learning to be a good leader. However, the self-fulfilling prophecy begins to take shape in the recruits' minds as they are told they don't have the ability to see the big picture. They are left with the impression that their only talent lies in pushing a broom, cleaning a toilet bowl, or doing what they are told. In this case, by giving the new recruit messages that he/she can't accomplish a given task, they begin to believe they can't do it either. So, we get exactly what we've reinforced, a level of work output that doesn't tap the full potential of the recruit!

*A common self-fulfilling prophecy for fire officers is that we don't expect enough of our people, so that's exactly what we get!*

## Pygmalion in Action

I'm sure you have other examples of Pygmalion in action in your department. Sometimes the negative self-fulfilling prophecy moves from the firefighter upward to the officer. There may be a general feeling among some of the firefighters that the officers or chief can't do the job, he'll just never be able to make it work. Just as expected, he can't seem to do the job. In our minds we condemn them to failure, and sure enough they fail. Could it be that when we believe someone will fail, we give them subtle negative messages or fail to support them and their programs?

Sometimes negative self-fulfilling prophecy works downward. As a supervisor, we can look at individuals below us and make predictions like, "Hell just never work out," or "She can't learn that." Guess what happens, you're right! They never seem to learn or they can't do the job. Why? We have created a threshold or limit we expect them to achieve. They only rise to the level of our expectations.

There are other examples of Pygmalion in the fire service. I've seen entire programs destroyed by individual attitudes about the potential success of the program. Several times I have seen ideas die at the staff table with the simple statement, "Well personally, I don't think that will work." Murphy's Law is based on that kind of pessimism. If you think things are going to fail there's little doubt they will not fail.

## Negativity — Who Loses?

The answer is, everyone loses. When ideas die, programs fail, and people are unsuccessful the profession becomes static or moves backward. The lifeblood of an organization is its ability to maintain momentum in the heavy current of change. In many cases, the fire service is literally swimming upriver against a torrent of pressures created by tax rebellion, community attitudes, budget cuts, and rejection of the "good guy" image of the fire service.

Pygmalion does not have to be negative. If we think things will work out for the best, in the end they just might! In mythology Pygmalion is remembered as the prince who got his wish. He is often pictured with a smile on his face.

Is there a smile on your face? The next time you get up in the morning take a look in the mirror. What do you see? Have you become a person who has preconceived notions of how things will work out? Do you act in a manner that assures the prophecies will be fulfilled? Considering the uncertain future the fire service seems to be facing, we can't afford negative Pygmalions. If we are going to effect programs, motivate people, embrace new ideas, and be innovative then we have to assume a positive, optimistic attitude.

Ask yourself this question, "If I am a Pygmalion, will I be able to live with my prophecies in the future? Am I thinking positively or negatively about myself and the future of the fire service?" The future will reveal whether the positive prophecies outweigh the negative ones in our attempt to deliver fire protection to our communities.

*Have you become a person who has preconceived notions of how things are going to work out?*

212

## COMMAND PRESENCE

Cliches are great! Sometimes we use short, succinct phrases to express something that is much more complex and sophisticated. For example, the hippies used to say "be cool." Trying to be cool is very complex. Another cliche was "looking good." This two word phrase was used to infer that a person "had their act together."

Every fire officer in a situation where an emergency is unfolding has to be cool, while looking good, and get his/her act together! We have a term for this. You will seldom find it in the definitions of Tactics and Strategy. It is not listed in acronyms, such as "R-E-C-E-O" OR "R-E-V-A-S," that offer suggestions on how to control the thought processes during an emergency. Nevertheless, the term we use is ***command presence.*** Command presence is easily described, but extremely difficult to achieve.

### What is Command Presence?

Command presence is an outward exhibition of total self-control in a situation that is recognized as an emergency. It is probably easier to describe the lack of command presence than it is to identify its existence. We have all seen fire officers under stress exhibit changes in personality, which creates stress in others. Some of the symptoms are talking loud, shouting orders, and the use of profanity to place emphasis upon a certain task. You may also observe dysfunctional behavior, such as running back and forth without accomplishing a given task. When the person in charge of a fire company lacks command presence, they lose control of the emergency at hand.

Few individuals who achieve the rank of fire officer immediately develop the ability to maintain command presence. This is not to say that they panic under pressure. As the flood of adrenalin accelerates our senses, anxiety often quickly follows. Even the most seasoned individuals experience this periodically.

We can look outside of the fire service and see examples of how command presence is built into individuals with tremendous responsibility. A classic example is the airline industry. It is almost eerie to listen to a tape recording of an aircraft in distress. Even when the aircraft suffers a severe problem the pilot maintains self-control. He/she clearly describes the conditions and explains the anticipated course of action to remedy the situation.

One of my friends refers to airline pilots as the deans of "cool school." Pilots are able to maintain command presence because they realize the only way the passengers and aircraft will survive is if he/she maintains control of the situation. This discipline has been bred into pilots through intensive training, education, and experience.

*Command presence is an outward exhibition of total self-control in a situation that is recognized as an emergency.*

## Develop Control Through Discipline

The key word is ***discipline.*** Used in this context, discipline does not imply punishment. The word discipline is derived from disciple, which means to be a student. A pilot is disciplined in an emergency situation because he has been a student of such conditions for a long time. He does not feel he is part of the problem, but rather sees himself as the only solution.

The same thing applies in the fire service. An emergency is something happening to an individual or group, not to you. The condition will continue to deteriorate unless someone takes control and provides a remedy to the situation. The fire service is one of the few public safety agencies that responds to emergencies while they are still in progress. Many other public safety services respond to emergencies that are over and only require a report to be filed.

One difference between law enforcement and fire protection is that in the vast majority of incidents a police officer responds to, the situation has already unfolded. In many cases the perpetrator has left the scene. Many of the emergencies that law enforcement responds to involve collecting data, rather than finding a remedy. When a police officer does have an emergency, it is often life threatening, dramatic,

*The emergency will continue to deteriorate unless someone takes control and provides a remedy to the situation.*

and highly visible. However, it is a minority of their activity.

On the other hand, in the fire service a significant portion of our emergencies are occurring at the time we respond. Our arrival and response time are critical to the outcome of the emergency in terms of destruction of life and property. This contributes to the fire officer's challenge to maintain command presence. In an emergency, we usually have resources available to get the crises under control in seven to ten minutes. Approximately 90% of our emergency responses fall in this category. We grow accustomed to a decision making process in which we quickly identify a needed outcome and develop a course of action.

At times we face an emergency that requires resources beyond those immediately available. When this stress occurs, the mind tends to shift into high gear. The mind attempts to quickly process all incoming information and make crucial decisions of how to deal with the crises with limited resources. The emergency has placed you in a position of making short-term, critical decisions with long-term impacts. It is essential that your mind be disciplined to handle this crises in advance through training and education.

## You Are The Solution

Early in my career, I was taught a valuable lesson along these lines. I was attending a school in a neighboring city that was taught by one of their senior fire officers. I was a young fire officer with limited experience with fully involved structures. On one of the class days, this senior officer was called out to re-

spond to a working structure fire. I chose to ride along with him to see how this individual dealt with the fire.

We could see the header a long way off. Even though I had no responsibility for the incident my adrenalin started pumping, my heart rate went up, and all of my "fight instincts" came into play. The senior officer I was with looked as though he was driving to pick up his lunch. When we arrived, I jumped out of the vehicle and made four or five hasty laps around the building to see what I could find out. As I looked back, I noticed the senior officer standing next to his vehicle studying the problem. His powers of observation were difficult to measure, but I noticed he appeared steady as a rock. Slowly he put on his turnout coat and helmet. Calmly and with a well thought out plan of attack, he directed incoming apparatus over the radio.

When the fire was over I asked him, "How do you stay so cool when all of that is going on?" His answer was classic, "Well Ron," he said, "you have to remember, I didn't start that fire. My job is to make sure that it goes out."

Our job is to make sure that emergencies are handled in the most effective manner possible. It is vital to the professional development of the fire service that the individuals in charge of emergency incidents see themselves as **the solution.** Command presence should be as essential to a good fire officer as apparatus, manpower, and programs. Command presence must be firmly established in the mind of each officer through training and discipline prior to the bell ringing to initiate an emergency response.

## Training and Education Develops Command Presence

There are some that believe command presence goes with the job. However, pinning a badge on a person's chest and calling them Lieutenant, Captain, or Battalion Chief does not result in the behavior we have discussed. There are specific courses of action that fire officers should engage in if they want to be assured they will develop command presence.

First and foremost is the **training and education** an officer receives prior to assuming the role. It is logical that an individual will feel more in control when they are knowledgeable about the short-term and long-term impacts of the decisions they make. The more knowledgeable a person is, the more likely he/she will maintain control. It is just as important for a fire officer to understand the anatomy of an emergency as it is for a doctor to understand the anatomy of a human being, if the officer is to correctly bring the emergency under control.

## Maintain Good Physical Conditioning

The second aspect of command presence is to **maintain good physical condition.** Like well trained athletes, individuals will perform better under stress when their bodies are capable of absorbing the physical aspects of stress with a minimum amount of impact. The function of decision making is mostly a cerebral exercise. However, if a person is not in good physical condition they can be more easily debilitated physically under stress, which causes the mental process to deteriorate. There is a term for this, **neurosthenia.** Neurosthenia is when an individual, being exposed to stress for a period of time, enters into a fatigued state. Decisions become improper and inaccurate.

## Evaluate Yourself And Others

The third way to maintain command presence is to **spend time critically evaluating your conduct at prior emergency scenes**. If your organization has the capability of tape recording emergency responses, get copies of the tapes where you have been the primary decision maker. Evaluate how you sound to others. The voice paints quite a picture under stress. Often a person's voice will become high pitched or shakey during an emergency.

One officer that I knew noticed during a simulator exercise that he had a habit of keying the radio microphone, followed by a long pause, and punctuated by uhhhh! He never realized what he was doing. This habit caused those istening to him over the radio to feel he was indecisive and confused about his course of action. A simple technique eliminated his problem. He would only place his index finger on the microphone when he had completed his thought process and was ready to talk. Then he would tap the mike with his index finger and key it as he began his message.

Another technique for improving your command presence is to **study individuals that exhibit command presence.** If you want to know how to be in control, study someone who is effective in maintaining control. This is another use of role modeling. Patterning your behavior after another person is not simply copying them, rather it is a shortcut to achieve your own personal style.



## Mentally Work Through Emergencies

Lastly, one of the best techniques to improve command presence is to **mentally engage in emergency situations.** Work through how you would handle the crises. I call this the Walter Mitty approach. Walter Mitty, as some of you will recall, was an individual whose life was a series of fantasies. The idea is to play mental games, thinking through how you would like to handle emergencies. Daydream your way through various scenarios.

One of the best ways of disciplining yourself to be ready to handle high stress circumstances is to do a form of dress rehearsal in your mind. This technique is an important part of the training of Olympic and professional athletes.

In summary, command presence is difficult to define. It cannot be learned in a classroom. Instead, it is part of the professional development of a fire officer. The lack of command presence erodes the credibility of the fire service. Command presence establishes confidence and credibility for the fire service profession and its officers.

# POST-PROGRAM ACTIVITY 1

## Developing Your Roadmap To Achievement

Throughout this course we have talked about a myriad of concepts, ideas, and techniques which we feel will help you to become a better fire officer. The next step in this course is the last, and possibly the most important — that of developing your personal roadmap to achievement.

Have you ever wondered why it seemed easier to have big dreams as a child than it is as an adult? Maybe it is because time, money, and circumstances were of no concern to us as children. Becoming an astronaut or the President of the United States seemed entirely feasible. As we grow older obstacles, failures, and other setbacks begin to loom in front of us. Dreaming gets replaced by just trying to get through the week. Yet, dreaming is the key to developing your Roadmap to Achievement.

Ask yourself two questions.

- what are my professional goals?
- what would my professional goals be if TIME, MONEY, and CIRCUM-STANCES were of no concern?

Is the answer to each of these questions different? For most of us the answers are vastly different. Many people ask, "Time, money, and circumstances are of major concern for me, so what's the sense of all this dreaming?" This answer is simple!

Everyone knows that in goal setting we are usually taught to make short term goals of one day to six months, then intermediate goals of six months to one year, and finally long term goals of one to five years or more. Most people "try" goal setting and get discouraged. The reality of obstacles such as time, money, and circumstances seem to constantly "rain on our parade."

The success stories of people we read about usually have two common threads. They are:

- the person had a dream and developed their short, intermediate, and long term goals around that dream.
- they kept moving toward the dream with presistence.

One of my favorite quotes talks about the importance of persistence.

> Nothing in the world can take the place of
> persistence. Talent will not; nothing is more
> common than unsuccessful men with talent.
> Genius will not; unrewarded genius is almost
> a proverb. Education will not; the world is
> full of educated derelicts. Persistence and
> determination alone are omnipotent. The
> slogan 'Press On' has solved — and always
> will solve — the problems of the human race.
> Calvin Coolidge

Remember, your ability to be persistent must be linked tightly to a dream or vision of where you want to go. The following exercise is a simple, but effective, way to begin to redevelop that childhood ability to dream big dreams.

## The Mental Helicopter Flight

Set aside at least one hour for this exercise. It is imperative that you choose a place that will be quiet with no interruptions. Have a blank pad and a pencil with you. If you like, you can softly play a tape of your favorite mood music. Get comfortable, turn on the music, close your eyes, and picture yourself climbing into a helicopter.

The helicopter takes off and you begin to enjoy the view. Suddenly you slow down and the helicopter begins to hover. Below is your perfect world five years from now. Time, money, and circumstances have not been obstacles for you in the last five years. How old are you? Where are you living? Are you in the fire service? What is your position? How have you improved as a professional from where you are today?

The keys to this exercise are:

- *don't be judgemental!* Even if your "perfect world" may seem totally un-realistic, don't judge it or stifle your thoughts.
- write down "what you saw" from the helicopter.

The final part of this exercise is to determine what can you begin doing **now** that will move you toward your perfect world five years from now. For example, if you found yourself as the fire chief of Los Angeles County in your perfect world five years from now, ask yourself, "What do I need to do to accomplish this goal, or one close to it?" You may find that you need to complete your college degree to achieve your dream. If this is the case, then

218

break the ultimate goal down into realistic, achieveable goals that you can begin immediately.

You may establish short term goals of getting a copy of your transcript and checking into admission requirements at the college nearest you. Your intermediate goal may be to apply to this college by the fall semester. Your long term goal may be to complete your degree within three years. At the same time you can set short, intermediate, and long term goals that you should be taking professionally to be moving in the direction of achieving the rank of chief within a period of time.

The last step is to establish a time when you will periodically review your Roadmap to Achievement. If your plan is to succeed it will have to be flexible and dynamic. Your professional goals may change, and with it so should your Roadmap. We have listed several books in the Reference section of this chapter (page 221) that you may find helpful as you strive to reach the goals you have set for yourself. We have stated a philosophy repeatedly in this course, "Don't search for excellence — create it!" The same holds true for you, **"Don't search for professionalism — create it!"**

# ROADMAP TO ACHIEVEMENT

*Instructions:* Begin by dreaming what your ultimate goal would be if time, money, and circumstances were not an issue. Then establish short, intermediate, and long term goals to help you achieve your ultimate goal!

**SHORT TERM GOALS**

**GOAL #1**_______________________

**TO ACCOMPLISH THIS I NEED TO:**

_____________________________

_____________________________

**GOAL #2**_______________________

**TO ACCOMPLISH THIS I NEED TO:**

_____________________________

_____________________________

**ULTIMATE GOAL**

_____________________________

_____________________________

**I WANT TO ACCOMPLISH THIS BY:**

_____________________________

**TO ACCOMPLISH THIS I NEED TO:**

_____________________________

_____________________________

**INTERMEDIATE GOALS**

**GOAL #1**_______________________

**TO ACCOMPLISH THIS I NEED TO:**

_____________________________

_____________________________

**GOAL #2**_______________________

**TO ACCOMPLISH THIS I NEED TO:**

_____________________________

_____________________________

**LONG TERM GOALS**

**GOAL #1**_______________________

**TO ACCOMPLISH THIS I NEED TO:**

_____________________________

_____________________________

**GOAL #2**_______________________

**TO ACCOMPLISH THIS I NEED TO:**

_____________________________

_____________________________

# REFERENCES — MODULE VIII


Dressler, Gary. *Organization & Management: A Contingency Approach.* Prentice Hall, 1976.

Drucker, Peter F. *The Effective Executive.* Harper & Row, 1966.

Lakein, Alan. *How to Get Control of Your Time and Your Life.* The New American Library, Inc., 1973.

Sayles, Leonard R. & Strauss, George. *Behavioral Strategies for Managers.* Prentice Hall, 1980.

Waitley, Denis. *Seeds of Greatness.* Pocket Books, A Division of Simon & Schuster, 1983.

# EPILOGUE

## LIGHTNING RODS, GARDEN RAKES, AND DOORMATS

As soon as man began to use symbols to express himself, he has been comparing himself to objects and animals in his surroundings. For example, a man who is strong was referred to as an ox. A cunning person is often referred to as a fox. To be cunning is also referred to as being "sharp as a razor." The list of analogies can go on and on.

We can use that same kind of an analogy when talking about our fire departments. For example, have you ever heard an organization referred to as "fat" or "lean?" We all do it. Analogies mean that the things we are comparing are "like" one another in a certain way. What about your organization, what is it like? If you were to compare your fire department to an animal or object, what image would you select? Most of us would like to select something that is positive. We would like an analogy to reflect a good image of ourselves and be an image that others would agree with.

Over the years I have developed three analogies that seem to represent the way fire departments are perceived by their communities. The three objects that I have selected for symbols are a:

- doormat
- garden rake
- and lighting rod.

### What Is A Doormat?

A *doormat* is an item that is necessary to keep the home tidy. It is placed at the point of entry and used to make sure that no dirt and debris is carried into the living area. Doormats lie just outside the safety of the home in all weather conditions.

A doormat serves a functional purpose for the home. People are generally unaware of its presence until our turbulent environment potentially threatens the interior of the home.

The doormat can be an analogy to today's successful fire department. It is the organization that lies unnoticed in the community through all "weather" conditions. It stands ready to serve all who enter through the doorways of the community. It is the organization that is often walked upon, but still lies ready to serve its occupants. The doormat fire department is ready to serve, especially when the turbulent environment threatens the interior of the community.

Once the crisis is over the doormat fire department fades into the background and receives little or no attention until the next crisis.

### What About A Garden Rake?

*Garden rakes* are tools we use to keep our lawns and gardens pleasing to the eye. There are at least a dozen varieties of rakes. Some are small and others are large; some are cheap and others are expensive.

Periodically, our trusty garden rake may strike out at us if we have inadvertently left it lying face-up in the garden. Should you happen to step on the rake, it may come swishing off the ground and crack you alongside the nose. When this happens, we often forget that it is our fault for leaving the rake there. Instead, we are likely to throw the rake out of sight in our anger. No matter how beneficial that rake was in keeping the landscape clean, it went from a tool to an enemy in a split second.

Sometimes our fire departments resemble garden rakes. They, too, are efficient and effective tools. They come in all sizes and shapes. Garden rake fire departments can also turn from a useful tool into an enemy in a split second.

Fire departments can behave in a disruptive and negative manner over such issues as labor unrest, management crisis, or if they believe they have been ignored or unappreciated by the community. Lack of meaningful dialogue or mutual respect between the fire department and the city administration can lead the garden rake fire department to unexpectedly and unwisely strike out.

Eventually the city administration may view the fire department as the enemy instead of a useful tool. Punishment can be as drastic as total removal of the fire department or finding a cheaper version, such as public safety.

## What Are Lightning Rods?

Most people don't even remember what a *lightning rod* is because modern society no longer requires them on individual homes. However, in some parts of the United States people find it necessary to place a piece of metal on top of high-rise buildings. This metal attracts electricity, or lightning, in the atmosphere and causes it to be diverted before it strikes the structure. This metal is a lightning rod. Lightning rods are higher and more visible than any other part of the building.

The lighting rod provides safety for the structure on which it is placed. Its purpose is to be visible and protect. It is designed to withstand even the most turbulent environment and still protect the structure.

Lightning rod fire departments are those which take the highest position in the community and provide total safety for the structures they protect. They are "doers" and are highly visible. They are the fire departments that get stronger as the environment becomes turbulent. However, they withstand this turblence and do not forget their primary mission is to provide fire protection and safety for the community. They can truly withstand and control even the greatest lightning bolt that may strike the community they protect.

## What Does The Future Hold?

Peter Drucker has said that we will be managing in turbulent times. I would hope that all the fire departments across this great nation would become a strong, lightning rod to their communities. Lightning rod fire departments place themselves in the midst of all the turbulence. They attract energy and attention because they are responsive to the community's fire problem.

Lightning rod fire departments are surrounded by flashes of lightning and claps of thunder. They are busy protecting and meeting the needs of the community. A lightning rod fire department is one that promotes new programs. It introduces changes that will achieve the basic objective of protecting life and property. It's a fire department that is promoting instead of defending. It is innovating instead of institutionalizing.

## What Kind of an Individual Are You?

The analogies that have been drawn could easily be related to individuals. In fact, the personality and make-up of the organization is a good reflection of the combined leadership styles of the individuals who are responsible for managing that organization.

What analogy would you draw for yourself? Perhaps you can come up with an analogy that better reflects how you see yourself or your fire department. Surely we have some bulldozers out there. We probably have a few life preservers. There have got to be some that can only be classified as "dynamite."

In the final analysis, however, it is not important what we think we are. It is important what others think we are. How would your community classify you and/or your fire department? We should all strive to develop a favorable image in our communities. You will know you have a favorable image in one of two ways. First, it can be seen by the degree of community support you will receive for your programs. Secondly, the confidence reflected in your department by those in political and administrative authority will indicate the community's image of your department.

We cannot afford to have a negative analogy applied to us. The fire service has an awesome responsibility. We need to be sure we use the right tools to do the job. If we do, our communities will be safer, our jobs will be more secure, and our communities will be stronger.

How does this relate to being a good fire officer? As you have progressed through this course, you have reflected on the type of person and officer you are or wish to become. Will you be a lightning rod fire department? What type of garden rake will you be? Will you be an effective, useful tool or one that strikes out and becomes useless? Do you serve as a trustworthy doormat that is always ready and willing to serve when the turbulent environment threatens? What type of officer would you like to be, how can you get there? It is our hope that this program will provide you with the concepts, skills, and vision to achieve the utmost in your role as a fire officer.

**Don't search for excellence — create it for yourself, your department, and for the fire service!**

# SUGGESTED FOR FURTHER READING


Bliss, Edwin C. *Getting Things Done.* Bantam, 1976.

Bennis, Warren & Nanus, Burt. *Leaders: The Strategies for Taking Charge.* Harper & Row, 1985.

Churchman, C. West. *The Systems Approach.* Dell Publishing, 1968.

Committee for Economic Development, Research and Policy Committee. *Improving Productivity in State and Local Government.* 1976.

Dressler, Gary. *Organization & Management: A Contingency Approach.* Prentice Hall, 1976.

Drucker, Peter F. *The Effective Executive.* Harper & Row, 1966.

Farber, Barry A. *Stress and Burnout in the Human Service Professions.* Pergamon Press, Inc., 1983.

Fournies, Ferdinand F. *Coaching for Improved Work Performance.* Van Nostrand Reinhold Co., 1978.

Gellerman, Saul W. *Management by Motivation.* American Management Association, Inc., 1968.

Harris, Thomas A., M.D. *I'm OK-You're OK.* Harper & Row, 1967.

Lakein, Alan. *How to Get Control of Your Time and Your Life.* The New American Library, Inc., 1973.

Lawrence, Gordon. *People Types & Tiger Stripes: A Practical Guide to Learning Styles, Second Edition.* Consulting Psychologists Press, 1982.

Lewis, David & Sharpe, Dr. Robert. *Thrive on Stress.* Warner, 1977.

Mackenzie, R. Alec. *The Time Trap.* McGraw-Hill, 1975.

Mager, Robert F. & Pipe, Peter. *Analyzing Performance Problems or 'You Really Oughta Wanna.'* Fearon Pitman Publishers, Inc., 1970.

McGregor, Douglas. *The Human Side of Enterprise.* McGraw-Hill, 1960.

McGregor, Douglas. *The Professional Manager.* McGraw-Hill, 1967.

Myers, Isabel Briggs. *Gifts Differing.* Consulting Psychologists Press, Inc., 1970.

Oncken, William Jr. *Managing Management Time.* Prentice Hall, 1984.

Ross, Joel E. *Managing Productivity.* Reston Publishing Co., 1977.

Sayles, Leonard R. & Strauss, George. *Behavioral Strategies for Managers.* Prentice Hall, 1980.

Selye, Hans. *Stress Without Distress.* The New American Library, 1975.

Theobald, Robert. *The Rapids of Change.* Knowledge Systems, 1987.

Waitley, Denis. *Seeds of Greatness.* Pocket Books, A Division of Simon & Schuster, 1983.

# Notes

# <u>Notes</u>

# Notes

# Notes

# Notes

# <u>Notes</u>

# <u>Notes</u>

<u>**Notes**</u>